Arbitration, Fairness and Stability:

Revenue Division in Collaborative Settings

Yair Zick
School of Physical and Mathematical Sciences
Nanyang Technological University
Singapore

Published by

AI Access

AI Access is a not-for-profit publisher with a highly respected scientific board that publishes open access monographs and collected works. Our text are available electronically for free and in hard copy at close to cost. We welcome proposals for new texts.

©Yair Zick 2015

ISBN 978-1-326-34119-0

AI Access
Managing editor: Toby Walsh
Monograph editor: Kristian Kersting
Collected works editor: Pascal Poupart
URL: aiaccess.org

Contents

I	Preface	i
II	Acknowledgments	ii
III	Abstract	iii

1 Introduction — 1
- 1.1 Overview of the Thesis — 1
- 1.2 Related Work — 2
- 1.3 Preliminaries — 5
  - 1.3.1 Classic TU Cooperative Games — 5
  - 1.3.2 OCF Games — 5
  - 1.3.3 Payoff Division — 8
  - 1.3.4 A Note on Computational Complexity — 10

2 Arbitration Functions and Stability in OCF Games — 13
- 2.1 The Arbitration Function — 13
  - 2.1.1 Properties of Arbitration Functions — 14
  - 2.1.2 Some Types of Arbitration Functions — 16
- 2.2 Redefining $\mathcal{A}$-Profitable Deviations — 18
- 2.3 The Arbitrated Core — 20
  - 2.3.1 Some Arbitrated Cores — 22
    - The Conservative Core — 22
    - The Sensitive Core — 22
    - The Refined Core — 22
    - The Optimistic Core — 23
- 2.4 Non-emptiness of Arbitrated Cores — 23
  - 2.4.1 The Conservative Core — 24
    - Convexity in OCF Games Revisited — 32
  - 2.4.2 The Sensitive Core — 37
  - 2.4.3 The Refined Core — 38

3 Computing Stable Outcomes in OCF Games — 45
- 3.1 Discrete OCF Games — 46
- 3.2 Computing Optimal Coalition Structures in OCF Games — 47
  - 3.2.1 Finding an optimal coalition structure — 48
  - 3.2.2 Limiting Interactions in OCF Games — 50
- 3.3 Computing Optimal Deviations — 52

    3.4 Computing $\mathcal{A}$-Stable Outcomes 58
    3.5 Beyond Tree Interactions 60
    3.6 Linear Bottleneck Games and the Optimistic Core 63
        3.6.1 Some Examples 65
        3.6.2 Computing Stable Outcomes in LBGs 66

4 Alternative Solution Concepts in OCF Games   71
    4.1 The Arbitrated Nucleolus 72
        4.1.1 Non-Emptiness of the Nucleolus 73
        4.1.2 Properties of the Nucleolus 74
    4.2 The Arbitrated Bargaining Set 76
        4.2.1 Non-Emptiness of the Bargaining Set 78
    4.3 The Shapley Value in OCF Games 82

5 Iterated Revenue Sharing   89
    5.1 Preliminaries 92
        5.1.1 Accounting for Individual Utilities 93
    5.2 Optimization and Regret in Homogeneous Functions 94
        5.2.1 Non-Differentiable Utility Functions 100
        5.2.2 Truthful Contracts 102
    5.3 Discounted Returns 103
    5.4 Applications 105
        5.4.1 CES Production Functions 105
        5.4.2 Cobb-Douglas Production Functions 106
        5.4.3 Leontief Functions 107
        5.4.4 Network Flow Games 108

6 Conclusions   111
    6.1 Future Work 112

I Preface

Before we get to the mathematical notation and formal definitions, I would like describe my PhD experience in a few words. It seems wrong to start this thesis with "let N be a finite set of players".

I started my PhD studies not knowing exactly what I wanted to work on. I was broadly interested in game theory and algorithms, but was not entirely sure what were "good" problems to work on. In retrospect, my initial interests could be seen as revenue division in portfolio selection problems, which, happily enough, I got to work on towards the end of my PhD (see Chapter 5). With the advice of Edith, I started reading a paper on cooperative games with overlapping coalitions [Chalkiadakis et al., 2010]. The problem it described was both natural and interesting: a group of people shared efforts in order to make money – how should they divide it among themselves? While working on the problem and developing the arbitration functions model, I came to realize that many of the interesting questions underlying the fields of AI and multiagent systems are in their core moral issues: how should money and power be divided? How would one define a "good" state of affairs? How does one fairly consider the opinions and preferences of several parties? How does one ensure that people are telling the truth?

The last four years of my life were the most intensive time I have ever experienced. During these four years I have moved countries, raised a family, and worked towards completing my PhD. I had a chance to meet some amazing people and make new friends, both within my professional community and outside it.

As a graduate student, you spend most of your time feeling that you don't know much. You are always working on what seems to be an unsolvable problem; finding a solution to this one problem often consumes you, and often enough actually finding the solution is quite anti-climatic. It was almost always the case that the solution is either trivial, or leads to a host of other, equally unsolvable problems. I suppose that this is a good thing: this is really what research is all about, a constant creative process, generating ideas that are put to paper and hopefully published.

The Jewish faith asserts that God created the world in words; in a sense, the act of creation is an act of enunciation. This is why many observant Jews believe that words and letters hold great power, power that can even rival that of God himself. The tale of the Golem of Prague is a classic example of the power of words over the physical world. According to legend, in order to protect his community from antisemetic attacks, the great rabbi Judah Loew ben Bezalel (the Maharal) creates a man made of clay, called a Golem. The rabbi breathed life to the Golem with prayers and rituals, concluding his act of creation by writing the word *Emeth* (truth, in Hebrew) on the Golem's forehead, at which point it comes to life, doing the bidding of its creator. The rabbi's creation is a formidable creature: it is stronger then any man, impervious to pain, and unfailingly loyal to its master. A tireless, powerful servant, the Golem sets about helping the Jewish community, fending off attacks on the Jewish quarter, and leading the people to prosperity and a life of ease. As most legends will, the story of the Golem comes to a tragic end: rabbi Loew loses control over his creation. The reason why is unclear: some accounts say that it is due to the Golem being put to work on the Shabbat (the Jewish day of rest), others that the Golem fell in love. In the end, rabbi Loew must destroy his creation; he erases the first letter of the word Emeth from the Golem's forehead, which then spells Meth (dead). The now inanimate Golem is then placed in a casket, and left in the attic of the Synagogue of Prague, where it (allegedly) rests to this day.

The creation of the Golem mirrors the creation of Man as described in the book of Genesis. Both Man and the Golem are fashioned of earth, and by the power of words, take on a life of their own.

The general meaning of the word Golem is inanimate or unformed; a Golem is something that is not yet alive, but has the potential to be so in time. I like to think that the Golem is an *idea*; when one writes down ideas, the very act of putting them to paper mimics the act of creation. An idea left unwritten and unshared is a Golem in the attic: useless, effectively dead.

While the writing process itself is extremely personal, it is not an individual feat. In order to come to life, an idea must be shared; otherwise, it may well not be. No matter how great your ideas are, they mean nothing unless they are shared by others. This is something that I have learned well through both my academic and family experiences; I suppose that this is really what motivates most of my work: finding good ways of sharing, and methods that ensure that people collaborate. I hope you enjoy reading about my little Golem.

II Acknowledgments

As is appropriate for a PhD thesis whose main object of study is collaboration, I have many people to thank. First, I would like to warmly thank my advisor, Edith Elkind. Edith has been the best advisor I could have hoped for. She possesses the unique talent of pointing out the right way to go, while giving me the freedom to choose my own path. Edith went beyond helping me conduct research; she made sure that I have the best start I can in the academic world, pushing me to interact with the academic community, meet collaborators and position myself as a researcher. She made me believe that I can do significant and interesting work, and that, for me, is the most important bit of all. Her mathematical insight, as well as her understanding of the academic world, have provided me with invaluable lessons that I will carry with me for the rest of my academic career.

I would like to thank my colleagues at NTU, especially my fellow graduate students and post-docs, with whom I have had many useful discussions. I had the chance to work with many interesting and exceptionally smart collaborators. I would like to thank Yoram Bachrach, George Chalkiadakis, Yuval Filmus, Nick Jennings, Ian Kash, Peter Key, Vangelis Markakis, Reshef Meir, Svetlana Obraztsova, Joel Oren, Dima Pasechnik, Maria Polukarov, Jeffery S. Rosenschein, and Alexander Skopalik; with your help, I have had a chance to work on the most rewarding and interesting problems.

I would like to thank Jeffery S. Rosenschein for hosting me at the Hebrew university of Jerusalem, to Nicolas R. Jennings for hosting me at the university of Southampton, and to Yoram Bachrach, Ian Kash and Peter Key for hosting me at Microsoft Research.

My graduate studies were generously funded by the Singapore International Graduate Award (SINGA) program, awarded by the Agency for Science, Technology and Research (A*STAR), Singapore.

I would like to thank my parents, Varda and Yehiel, and my siblings, Aviad and Anat, who supported me from faraway Israel during my time in Singapore. I would also like to thank my late grandmothers, Yehudit and Sarah, who are always with me in spirit, and whose legacy I hope to upkeep.

Finally, I would like to thank my wife, Jolene, for her love and support throughout these often trying times; you have been an inspiration and a source of sanity. I dedicate this thesis to my two sons, Noam and Ayal, with much love.

III Abstract

We study the problem of dividing revenue among several collaborative entities. In our setting, each player possesses a finite amount of some divisible resource, and may allocate parts of his resource in order to work on various projects. Having completed tasks and generated profits, players must agree on some way of dividing profits among them. Using the overlapping coalition formation (OCF) model proposed by Chalkiadakis et al. [2010] as the basis of our work, we develop a model for handling deviation in OCF games. Using our framework, which we term *arbitration functions*, we propose several new solution concepts for OCF games. In the first part of this thesis, we analyze the *arbitrated core* of an OCF game; we show some necessary and sufficient conditions for the non-emptiness of some arbitrated cores, and explore methods for computing outcomes in the core of an OCF game. Next, we describe and analyze the arbitrated nucleolus, bargaining set and two notions of a value for OCF games. All of the solution concepts we propose draw strong similarities to their non-OCF counterparts, and in fact contain the classic cooperative solution concepts as a special case. We conclude this thesis by proposing a solution concept for OCF (and non-OCF) cooperative settings that is based on a natural revenue allocation dynamic. In our setting, player revenue in time t acts as his available resources at time $t+1$. Assuming that players are not myopic and care about their long-term rewards, we show that under certain conditions, players' incentives become aligned with what is socially optimal. In other words, choosing a payoff division that maximizes long-term social welfare will be agreeable for all players.

Chapter 1

Introduction

A group of players collaborates, sharing resources in order to complete revenue generating tasks; how should profits be divided? In this thesis we try and provide several answers to this question.

1.1 Overview of the Thesis

In cooperative games with overlapping coalitions (overlapping coalition formation games, or OCF games) [Chalkiadakis et al., 2010], each player controls a divisible resource (e.g. time, money, a commodity). A coalition of players is formed by each player contributing some part of his resource to working with others, generating profits. One simple example of such a setting is a production market, where each player possesses a given quantity of some commodity (say flour, sugar, oil); groups of players can join in order to produce some product of value (say, by allocating 50kg of flour, 40kg of sugar and 30kg of oil, one can produce cookies).

After players form coalitions and generate revenue, the members of each coalition may distribute that coalition's revenue among themselves in any way they wish. This thesis studies properties of some classes of revenue divisions. Defining what constitutes a "good" payoff division is a complex task in itself; in the OCF setting, it is made even more so by the intricate nature of deviation in this setting.

We begin by describing a new formal model for handling deviation in cooperative games with overlapping coalitions. Chapter 2 describes *arbitration functions*, a general method for describing how players react when some group of players decides to deviate from a proposed division of labor and profits. Using arbitration functions, we describe a new solution concept for OCF games, the *arbitrated core*, and show some of its general properties. We then provide an alternative characterization of the core for various arbitration functions, showing a connection to classic cooperative games, as well as some conditions that ensure that the core is not empty (under different reactions to deviation).

In Chapter 3, we analyze the computational complexity of finding payoff divisions in the core of an OCF game. We show that while finding stable outcomes is hard —even for very limited subclasses of OCF games— it is possible to find core allocations in polynomial time if one makes certain assumptions on the structure of player interaction, and on how they react to deviation. We then show that the core of a class of OCF games called *linear bottleneck games* has a non-empty core, even when one assumes that reaction to deviation is extremely lenient. Moreover, finding core allocations can be done in polynomial time using linear programming techniques.

In Chapter 4, we analyze alternative solution concepts for OCF games. We define OCF analogues of three well-known solution concepts: the nucleolus, the bargaining set and the Shapley value. We show that our solution concepts maintain many of the properties of their non-OCF counterparts; moreover, we show that the definitions we provide are a generalization of the classic solution concepts, under an appropriate arbitration function.

In Chapter 5, we offer an alternative method of dividing profits in collaborative settings which is based on a natural dynamic. Rather than assuming that the game played is a one-shot interaction, we assume that the game is played in rounds, where the profits that each player receives in a given round act as his resources for the next round. We show that under this dynamic, players are incentivized to consider not only their long-term profits, but those of others. In particular, we show that when certain conditions on the utility function hold, players are happy with the socially optimal allocation, i.e. what is best for society is also best for individuals. We study the effect of discounted returns on our model, showing that if players are sufficiently far-sighted, they would still be inclined to do what is socially optimal, as this is what is optimal for them as well in the long run. Finally, we show that our results imply that players are also truthful: if they have private information to report that may affect the profit division, they gain nothing by misreporting if their incentives are aligned with what is socially optimal.

1.2 Related Work

The OCF model applies to settings where rational agents divide their resources among several projects, receiving profits from each project they contributed to. This scenario occurs in several settings. For example, as shown in Section 3.6, multi-commodity network flows can be easily modeled as an OCF game, as well as several other fractional optimization problems. For example, a fractional matching problem can be thought of as an OCF game, where each partial coalition between a pair of players describes the amount of resources that each of them assigns that task, and the value that is generated by them joining forces must be divided between them in some manner. There exists extensive literature on the optimal coalition structure generation problem in cooperative settings with overlapping coalitions. Shehory and Kraus [1996] have initiated this line of work; in their work, they describe a distributed coalition formation algorithm, where players may divide their resources between several tasks in order to achieve better outcomes. Dang et al. [2006] consider overlapping coalitions in sensor networks, where the players are sensors that are tasked with tracking objects. Lin and Hu

[2007] explore overlapping coalition formation using a parallel algorithm. In all these works, the underlying assumption is that agents are fully cooperative, seeking to maximize social welfare, and are not interested in maximizing their own gains. In our setting, we explore game-theoretic aspects of strategic behavior in cooperative settings, taking agent incentives into account.

Other models where overlapping coalitions arise are collaboration networks. Consider a social network such that each player is a weighted node, and there is some value for assigning a certain portion of an player's weight to collaborating with his neighbors. Such settings have received much attention in recent years (see Jackson [2003] for a survey). More specifically, Anshelevich and Hoefer [2010] discuss strategic aspects of network formation in settings where players may participate in more than one coalition; their analysis focuses more on individual rationality and pairwise equilibria in such networks, rather than group stability. More recently, Ackerman and Brânzei [2012] analyze research collaboration using a graphical model, where authors devote some of their efforts to collaborative projects, and receive some credit for their work based on the authorship order model. Like the previous work mentioned, their work focuses on individually rational and pairwise stable outcomes, rather than group stability.

Computational aspects of cooperative games have been the subject of extensive research (see [Chalkiadakis et al., 2011] for an extensive literature review). Cooperative games with overlapping coalitions have initially been discussed in the literature by Aubin [1981], who coined the term *fuzzy games*. However, the OCF model that our work is based on is defined in Chalkiadakis et al. [2010]. While the two models share some of their definitions, the fundamental approach is quite different. The first, and arguably most important, difference is that Chalkiadakis et al. [2010] assume the formation of several partial coalitions, whereas Aubin [1981] assumes that all players pool their resources to work on a single task. Assuming the formation of a single coalition eliminates several of the subtleties that games with overlapping coalitions can exhibit, as was emphasized in [Chalkiadakis et al., 2010]. A key point that was raised by Chalkiadakis et al. [2010] is that several separate interactions are required in order to observe the more interesting behaviors exhibited in OCF games. More specifically, interesting behavior arises when non-deviators still interact with deviators post-deviation; this setting does not arise in fuzzy games.

In Chapter 3, we employ the notion of treewidth in order to obtain polynomial-time algorithms for computing stable outcomes in OCF games. Several works have successfully employed treewidth as a means for achieving computational tractability in cooperative games. Brafman et al. [2010] employ techniques that are rather similar to those used in Chapter 3 in order to compute stable outcomes in cooperative TU planning games; Greco et al. [2011] make direct use of Courcelle's theorem [Courcelle, 1990] in order to compute solution concepts for several classes of classic cooperative games that have a graphical structure; more recently, Voice et al. [2012] study the coalition structure generation problem for a class of games with a graphical structure, employing treewidth to obtain poly-time algorithms.

One can see the iterated revenue sharing approach taken in Chapter 5 as a model of dynamic coalition formation in cooperative games with overlapping coalitions. Such models have been explored in non-OCF cooperative games (see, e.g. Arnold and Schwalbe [2002], Lehrer and Scarsini [2013], Shehory and Kraus [1996]), but have either dealt with non-OCF notions of stability, or focused on finding optimal coalition structures. It would be interesting to see the role arbitration functions serve in decentralized dynamic formation of stable outcomes. One could use arbitration functions as a method of controlling the robustness of the coalition formation process. If an outcome at some stage of the coalition formation process is stable, more permissive reaction to deviation is allowed in future rounds; if the outcome is unstable, players can be assumed to be less tolerable towards deviators. Thus, by controlling the extent to which players are allowed to deviate, one can arrive at a stable outcome.

Several papers study dynamics of coalition formation among players (see, e.g. Sandholm et al. [1999], Shehory and Kraus [1996]; or a literature review in Chalkiadakis et al. [2011]) however, revenue division and its effects are not studied. Revenue division in dynamic settings has been studied from a variety of perspectives. There exist several works on dynamic coalition formation and iterated revenue division. For example, in the setting studied by Lehrer and Scarsini [2013], a sequence of allocations is proposed for a given function $v : 2^N \to \mathbb{R}$, and various dynamic solution concepts are proposed. However, in this setting, the value of a coalition does not depend on the revenue it receives, but is rather given by the function v. Elkind et al. [2013] study computational aspects of solution concepts in dynamic weighted voting games, where player weights and the quota change over time. While in this setting the value of coalitions changes as a function of player resources (namely, their weights), the way weights change is given exogenously as part of the problem input, while in the setting described in Chapter 5 we assume that the weight of a player at time t is given by the revenue sharing scheme that the players choose.

Chapter 5 uses individual regret to assess the desirability of resource allocations. Regret minimization is a well-known benchmark in learning [Even-dar et al., 2009, Zinkevich, 2003] (see [Nisan et al., 2007, Chapter 4] for an overview). In this setting, we are given a sequence of functions $f_t : \mathbb{R}^n \to \mathbb{R}$. Prior to observing f_t, an algorithm A chooses a vector $\mathbf{x}_t \in \mathbb{R}^n$ and receives a payoff of $f_t(\mathbf{x}_t)$. The regret experienced by A at time T measures its average performance against a static contract $\mathbf{x} \in \mathbb{R}^n$, i.e. the difference between $\frac{1}{T}\sum_{t=1}^{T} f_t(\mathbf{x}_t)$ and $\frac{1}{T}\min_\mathbf{x} \sum_{t=1}^{T} f_t(\mathbf{x})$. An algorithm A has low regret if this difference is $o(1)$. While we use a similar benchmark, our analysis and optimization objectives are quite different from that of the regret minimization literature.

The results in Chapter 5 are similar to the problem of *portfolio selection* [Cover, 1991]: given a set of n stocks and a sequence of changes to the stock values (given by vectors in $\mathbb{R}^n$), find an optimal investment portfolio, i.e. an investment strategy, that maximizes total wealth. There is a significant body of work analyzing the performance of algorithms for portfolio selection, and its connection to regret has been previously studied (see, e.g., Gofer and Mansour [2011] and the citations within). Our work differs from the portfolio selection problem in several ways. First, in portfolio selection problems, stocks are not rational agents: they are not interested in having more money invested in them; in our work, individual elements where revenue is invested care about their welfare. Second, uncertainty plays an important role in the portfolio selection problem, whereas our model is deterministic (although adding uncertainty is certainly an important direction for future work).

1.3 Preliminaries

We begin by describing some necessary notation and basic concepts that will be used in this thesis. Throughout this work, we write $N = \{1, \ldots, n\}$ to be a set of players. We use boldfaced lowercase letters to denote vectors, and uppercase letters to denote sets. Given a set $S \subseteq N$, we write $e^S \in \mathbb{R}^n_+$ as the indicator vector of S; that is, $e_i^S = 1$ if $i \in S$, and is 0 otherwise. Finally, given a vector $\mathbf{x} \in \mathbb{R}^n$ and some subset $S \subseteq N$, we write $x(S) = \sum_{i \in S} x^i$, and $\mathbf{x}|_S$ to be the vector $\mathbf{x}$ with all coordinates $i \notin S$ set to 0.

1.3.1 Classic TU Cooperative Games

This work deals with profit sharing and group incentives in collaborative settings; as such, its foundation lies in *cooperative game theory* (see Peleg and Sudhölter [2007] for a primer on cooperative games). We introduce some basic notations here. A *transferable utility (TU) cooperative game* is defined by a set of players $N = \{1, \ldots, n\}$ and a *characteristic function* $u : 2^N \to \mathbb{R}$ with $u(\emptyset) = 0$. The set of *feasible payoffs* for a game $\mathcal{G} = \langle N, u \rangle$ is the set of all vectors $\mathbf{x} \in \mathbb{R}^n_+$ such that $x(N) = u(N)$. It is sometimes assumed that players are allowed to form *coalition structures* [Aumann and Drèze, 1974]; a coalition structure CS is simply a partition of N, and the value of CS is the sum of the values of its constituent coalitions. If S is in CS, then the value of S, $u(S)$, can be freely divided among all members of S, but not given to any players that are not members of S.

A *solution concept* is a function that assigns every TU game $\mathcal{G} = \langle N, u \rangle$ a set of feasible payoff vectors. A solution concept that associates every game $\mathcal{G}$ with a single payoff vector is called a *value*. Solution concepts for TU games abound in the literature; for a detailed review see Peleg and Sudhölter [2007], and Brânzei et al. [2005].

1.3.2 OCF Games

Overlapping coalition formation games were first introduced by Chalkiadakis et al. [2010]. Given a set of players $N = \{1 \ldots n\}$, a *partial coalition* of players in N is a vector $\mathbf{c} \in [0, 1]^n$; each player i in N may contribute a fraction of their resources to $\mathbf{c}$, which is simply the i-th coordinate of $\mathbf{c}$. In what follows, we will omit the word "partial", and refer to vectors in $[0,1]^n$ as *coalitions*. Of particular interest are the players who are actively participating in $\mathbf{c}$; these are players who may rightfully claim a share of the profits from $\mathbf{c}$. Moreover, they are the players who may potentially get hurt if some player changes its contribution to $\mathbf{c}$. Formally, the *support of the coalition* $\mathbf{c}$ is the set $\{i \in N \mid c_i > 0\}$, and is denoted $supp(\mathbf{c})$. We now recall the formal definition of OCF games, given in [Chalkiadakis et al., 2010].

Definition 1.1. An *OCF game* $\mathcal{G} = \langle N, v \rangle$ is given by a set of players $N = \{1, \dots, n\}$ and a *characteristic function* $v : [0,1]^n \to \mathbb{R}$ assigning a real value to each partial coalition; we require that $v(0^n)$ is 0.

The characteristic function of an OCF game can be quite general; we do not even require that v is monotone. The reasoning behind this apparent leniency is that given v, the players in N may form several coalitions in order to optimize their revenue, a process which results in *coalition structures*. This definition does carry with it certain limitations. For example, we implicitly assume that if players invest the same efforts they will receive the same revenue; moreover, it assumes that there are no interdependencies between tasks i.e., the value of a coalition only depends on the amount of resources invested in it.

A *coalition structure* over N is a finite list of coalitions $CS = (\mathbf{c}_1, \dots, \mathbf{c}_m)$ such that $\sum_{j=1}^m \mathbf{c}_j \leq 1^n$. The coalition structure CS can be thought of as an $n \times m$ matrix whose rows sum to (at most) 1. The *cardinality* of CS is the number of coalitions in CS, and is denoted $|CS|$. We require that each row sums to no more than 1 so that CS is a valid division of players' resources, i.e., each player gives at most 100% of its resources to a coalition structure. For the sake of clarity, we use set notation when referring to coalition structures e.g., when referring to a coalition $\mathbf{c}$ which is listed in CS, we write $\mathbf{c} \in CS$; when referring to a coalition structure CS' which is a sublist of CS, we write $CS' \subseteq CS$. Coalition structures in classic cooperative games are partitions of the player set, i.e., a set of disjoint subsets of N whose union is N. In particular, it is impossible to have a set appearing twice in the same coalition structure. In OCF games, however, a coalition structure is a partition of player resources; therefore, it is possible for a coalition structure CS to have the same coalition appearing more than once. Having two identical coalitions in a coalition structure means that the players are completing two separate tasks, which require the same resources and generate the same revenue.

Example 1.2. Consider a setting where two researchers collaborate. They are able to either write a single high-quality research paper, which will require all of their time, and will generate a revenue of 100,000\$ (in grant money). Alternatively, if they split their time equally between two research projects, it will result in two papers of equal quality, that generate 70,000\$ each. In this setting, it is in the players' best interest to split into two identical coalitions of the form $\binom{0.5}{0.5}$, rather than one coalition.

The total revenue generated by the coalition structure CS is simply $\sum_{\mathbf{c} \in CS} v(\mathbf{c})$, and is referred to as $v(CS)$.

Given a set of players $S \subseteq N$, we say that the coalition structure CS is *over* S if for all $\mathbf{c} \in CS$ we have that $supp(\mathbf{c}) \subseteq S$. We denote the set of all coalition structures over S by $\mathcal{CS}(S)$. Given a coalition structure CS, we denote the *weight vector of* CS to be $\mathbf{w}(CS) = \sum_{\mathbf{c} \in CS} \mathbf{c}$. The vector $\mathbf{w}(CS)$ indicates the total amount of resources that each player in N has invested in the coalition structure CS. Note that $\mathbf{w}(CS) \in [0,1]^n$, and if $CS \in \mathcal{CS}(S)$, then $\mathbf{w}(CS) \leq \mathbf{e}^S$; we say that $CS \in \mathcal{CS}(S)$ is *efficient* if $\mathbf{w}(CS) = \mathbf{e}^S$. We also write $\mathbf{w}_S(CS)$ to denote the total weight of S in CS; namely, the i-th coordinate of $\mathbf{w}_S(CS)$ equals the i-th coordinate of $\mathbf{w}(CS)$ if i is in S, but it is 0 otherwise.

Recall that a function $f : \mathbb{R}^n \to \mathbb{R}$ is called superadditive if for all $\mathbf{c}, \mathbf{d} \in \mathbb{R}^n$ we have that

$$f(\mathbf{c}) + f(\mathbf{d}) \leq f(\mathbf{c} + \mathbf{d}).$$

In the case of OCF games, the characteristic function v is from $[0,1]^n$ to $\mathbb{R}$; therefore, we must also require that the vector $\mathbf{c} + \mathbf{d}$ is still in the domain of the function, i.e., that $\mathbf{c} + \mathbf{d} \in [0,1]^n$. In the non-OCF setting, a similar notion of superadditivity is often employed; a function $u : 2^N \to \mathbb{R}$ is called superadditive if for all sets $S, T \subseteq N$ such that $S \cap T = \emptyset$ we have

$$u(S) + u(T) \leq u(S \cup T);$$

this notion can be seen as a special case of superadditivity in OCF games, applied only to indicator vectors of sets.

As shown in Example 1.2, players may form coalition structures in order to increase revenue; such considerations naturally give rise to the following definition: *The superadditive cover of v* is defined as

$$v^*(\mathbf{c}) = \sup\{v(CS) \mid CS \in \mathcal{CS}(N), \mathbf{w}(CS) \leq \mathbf{c}\}.$$

The value $v^*(\mathbf{c})$ is the most that the players can make using the resources of $\mathbf{c}$. The function v^* is the superadditive cover of v in the sense that it is the minimal superadditive function from $[0,1]^n$ to $\mathbb{R}$ such that $v^*(\mathbf{c}) \geq v(\mathbf{c})$ for every $\mathbf{c}$. Note that a similar notion of superadditive covers exists for classic coalitional games [Aumann and Drèze, 1974]: given a function $u : 2^N \to \mathbb{R}$, its superadditive cover is the function $u^* : 2^N \to \mathbb{R}$ where

$$u^*(S) = \max\{\sum_{T \in \mathcal{P}} v(T) \mid \mathcal{P} \text{ is a partition of } S\}.$$

While we do not, in general, restrict the behavior of v, we do require that v has the following property.

Definition 1.3. We say that v has the *efficient coalition structure property* if for every $\mathbf{c} \in [0,1]^n$, there is some $CS \in \mathcal{CS}(N)$ such that $\mathbf{w}(CS) \leq \mathbf{c}$ and $v^*(\mathbf{c}) = v(CS)$.

For such functions, the sup in the definition of v^* can be replaced with a max. In this work, we assume that all OCF games have characteristic functions with the efficient coalition structure property. The efficient coalition structure property plays a crucial role in making the notion of deviation sensible. For example, consider the function $f(\mathbf{c}) = \alpha$ for all $\mathbf{c} \in [0,1]^n$ such that $\mathbf{c} \neq 0^n$, where α is some positive constant. The superadditive cover of f is $f^*(\mathbf{c}) = \infty$ for all $\mathbf{c} \neq 0^n$. In particular, this means that no coalition structure is optimal: it is always possible to increase social welfare by splitting some coalition into two non-zero coalitions.

To provide some intuition for the concept of OCF games, we now present a class of OCF games. It is often useful to think of the players as using their resources to complete a given set of tasks. Such games are described in Chalkiadakis et al. [2010] and are called *Threshold Task Games (TTGs)*. A TTG comprises of a finite list of tasks, $T = \{t_1, ..., t_k\}$; each t_l requires some weight $w(t_l) \geq 0$ for its completion, and gives a certain payoff $p(t_l) \geq 0$. Each player i has some weight $w_i \geq 0$ that he may allocate to the completion of any task. The worth of a coalition is

$$v(\mathbf{c}) = \max\{p(t_l) : w(t_l) \leq \sum_{i=1}^{n} c^i w_i\};$$

that is, the value of a coalition is simply the value of the highest paying task that the players can complete with their combined weight. Threshold Task Games are a natural extension of *Weighted Voting Games* [Maschler et al., 2013]; in a weighted voting game, each player $i \in N$ has a non-negative weight w_i; a set of players $S \subseteq N$ has a value of 1 if the combined weight of its members is at least some given non-negative quota q. Thus, weighted voting games are simply threshold task games with a single task with a weight of q and a payoff of 1.

1.3.3 Payoff Division

Having divided into coalitions and generated revenue, players must decide on how to divide the revenue amongst themselves in some agreeable manner. OCF games allow *transferable utility*: the value of any coalition $\mathbf{c}$ may be freely divided among its constituent players, i.e., $supp(\mathbf{c})$. First, we formally describe payoff divisions in OCF games.

An *imputation* for a coalition structure $CS = (\mathbf{c}_1, ..., \mathbf{c}_m) \in \mathcal{CS}(N)$ is a list of vectors $\mathbf{x} = (\mathbf{x}_1, ..., \mathbf{x}_m)$; for each $1 \leq j \leq m$, $\mathbf{x}_j$ is a vector in $\mathbb{R}^n$ that describes how much does each player in i receive from $\mathbf{c}_j$. Simply put, given that players formed $\mathbf{c}_j$ and generated a revenue of $v(\mathbf{c}_j)$, $\mathbf{x}_j$ specifies the way in which this revenue is divided. Given some $\mathbf{c} \in CS$ and an imputation $\mathbf{x}$, we refer to the way the value of $\mathbf{c}$ is divided as $\mathbf{x}(\mathbf{c})$; thus, the payoff to player i from coalition $\mathbf{c} \in CS$ under $\mathbf{x}$ is $x_i(\mathbf{c})$. In order for $\mathbf{x}$ to be a valid division of payoffs to players it must satisfy:

Individual Rationality: $\sum_{\mathbf{c} \in CS} x_i(\mathbf{c}) \geq v^*(\mathbf{e}^{\{i\}})$ for all $i \in N$.

Payoff Distribution: for all $\mathbf{c} \in CS$ we have $\sum_{i \in N} x_i(\mathbf{c}) \leq v(\mathbf{c})$, and if $c_i = 0$ then $x_i(\mathbf{c}) = 0$.

We denote the set of all imputations over CS as $I(CS)$; observe that intercoalitional transfers are not allowed. That is, if a player $i \in N$ is not in the support of $\mathbf{c}$, then i cannot expect to receive any payoff from $\mathbf{c}$. We call a tuple $(CS, \mathbf{x})$, where CS is a coalition structure and $\mathbf{x} \in I(CS)$, a *feasible outcome*. We denote by $\mathcal{F}(S)$ the set of all feasible outcomes over S. Thus, $\mathcal{F}(N)$ refers to all possible ways that players can form coalition structures and divide payoffs.

Let $(CS, \mathbf{x}) \in \mathcal{F}(N)$ be a feasible outcome, we define the *payoff* to a player $i \in N$ as $p_i(CS, \mathbf{x}) = \sum_{\mathbf{c} \in CS} x_i(\mathbf{c})$. This is the total payoff of i from all coalitions in CS. Similarly, the total payoff to a set S is denoted $p_S(CS, \mathbf{x}) = \sum_{i \in S} p_i(CS, \mathbf{x})$.

Definition 1.4. Given a set $S \subseteq N$ and $CS \in \mathcal{CS}(N)$ the *coalition structure CS reduced to S* is defined as

$$CS|_S = (\mathbf{c} \in CS : supp(\mathbf{c}) \subseteq S).$$

These are all coalitions that are comprised solely of members of S.

The coalitions in $CS|_S$ are those that the set S fully controls. Hence a deviation by the members of S would only affect non-deviators if S changes any coalitions outside of $CS|_S$. Thus, if the members of S are unhappy with the payments they receive at a given coalition structure, they may be interested in removing some (or all) of their support from coalitions in $CS \setminus CS|_S$. We are now ready to formally define deviations in OCF games.

Definition 1.5. Suppose that $CS \setminus CS|_S = (\mathbf{c}_{j_1}, \ldots, \mathbf{c}_{j_k})$. A coalition structure $CS' = (\mathbf{d}_1, \ldots, \mathbf{d}_k)$ is a *deviation of S from CS* if for all $1 \leq \ell \leq k$ we have that $\mathbf{d}_\ell \leq \mathbf{c}_{j_\ell}$ and $\mathbf{d}_\ell \leq \mathbf{e}^S$. Alternatively, given a coalition structure $CS \in \mathcal{CS}(N)$ and some set $S \subseteq N$, for all $\mathbf{c} \in CS \setminus CS|_S$ we write $\mathbf{d}_{CS'}(\mathbf{c}) \in CS'$ to be the deviation of S from $\mathbf{c}$, requiring that $\mathbf{d}_{CS'}(\mathbf{c}) \leq \mathbf{c}$ and $\mathbf{d}_{CS'}(\mathbf{c}) \leq \mathbf{e}^S$. Under this notation, $\mathbf{d}_{CS',i}(\mathbf{c})$ specifies the amount of resources that player i withdraws from the coalition $\mathbf{c}$ under CS'.

When the deviation CS' is understood from the context, we omit the subscript $\mathbf{d}_{CS'}(\mathbf{c})$ and refer to the deviation of S from $\mathbf{c}$ as $\mathbf{d}(\mathbf{c})$. The deviation CS' describes how a set S withdraws resources from some coalitions in $CS \setminus CS|_S$; we require that $\mathbf{d}(\mathbf{c}) \leq \mathbf{c}$ for all $\mathbf{c} \in CS \setminus CS|_S$ since S cannot withdraw more resources from the coalition $\mathbf{c}$ than it has invested in it; we require that $\mathbf{d}(\mathbf{c}) \leq \mathbf{e}^S$ since the deviation CS' must involve only members of S, thus for any $i \notin S$ we have $d_i(\mathbf{c}) = 0$.

Note that if S withdraws all of its resources from $CS \setminus CS|_S$, then the total weight available to S is $\mathbf{w}(CS|_S) + \mathbf{w}_S(CS \setminus CS|_S) = \mathbf{w}_S(CS)$.

Example 1.6. Consider a three player game described as follows. There are several types of tasks: t_{12} requires 50% of player 1's resources and all of player 2's, and gives a payoff of 15; t_{13} requires 50% of player 1's resources and all of player 3's, and gives a payoff of 20. Moreover, player 1 can work alone on task t_1, which requires 50% of his resources and provides a payoff of 5. Finally, there is a task T which can be completed by any player if he allocates 50% of his resources to it, and the payoff from T is 3. The optimal coalition structure in this case is

$$CS = \left(\mathbf{c}_1 = \begin{pmatrix} 0.5 \\ 1 \\ 0 \end{pmatrix}, \mathbf{c}_2 = \begin{pmatrix} 0.5 \\ 0 \\ 1 \end{pmatrix}\right).$$

The coalition $\mathbf{c}_1$ completes task t_{12} and $\mathbf{c}_2$ completes task t_{13}; thus, $v(\mathbf{c}_1) = 15$ and $v(\mathbf{c}_2) = 20$.

A deviation of player 1 from CS would be a coalition structure

$$CS_1 = \left(\begin{pmatrix} \beta \\ 0 \\ 0 \end{pmatrix}, \begin{pmatrix} \gamma \\ 0 \\ 0 \end{pmatrix}\right)$$

such that $\beta \leq 0.5$ and $\gamma \leq 0.5$. If $\beta = 0$, then player 2 is unhurt by the deviation; if $\gamma = 0$ then player 3 is unhurt by the deviation.

It is possible that player 1 deviates from one coalition while not changing another. Moreover, it is possible that a coalition from which resources were withdrawn can still generate some profit. In our case, if player 1 withdraws all resources from c_1, the coalition with player 2 alone generates a profit of 3 after the deviation. This opens opportunities for more nuanced player behavior post deviation: player 3 may decide not to break his agreement with player 1 if $\gamma = 0$, or if his current payment from his coalition is less than 4, in which case player 1 may still give him his original payment (which will, presumably, be agreeable).

As Example 1.6 illustrates, deviations in OCF games may leave some of the non-deviators unhurt by the deviation. Moreover, it may be possible for deviators to keep paying non-deviators the same amount that they received prior to deviation. If deviators agree to assume the marginal loss of their deviation, it is possible that non-deviators will agree to maintain coalitions with them. These notions are formalized and discussed in detail in Chapter 2.

1.3.4 A Note on Computational Complexity

We briefly recall some notions from computational complexity. A detailed overview of algorithms and complexity can be found in Cormen et al. [2001] and Garey and Johnson [1979]. Given a problem we wish to solve algorithmically, suppose that m is some natural parameter of the problem. One natural parameter is the number of bits required in order to represent the input x, denoted $\|x\|$; however, in the case of cooperative games, one of the most natural problem parameters would be n, the number of players. We note that a naive representation of a cooperative game is a list of 2^n coalitions, which is of exponential size; however, algorithms that run in time exponential in n are not considered efficient (see Chalkiadakis et al. [2011], and the discussion in Chapter 3 for further details).

We say that a problem instance runs in time polynomial in m when there is some fixed-degree polynomial $p(x)$ such that the number of steps taken by an algorithm is at most $p(m)$; alternatively, if the degree of p is k, we would say that the problem instance is in $\mathcal{O}(m^k)$. We stress that k cannot depend on the parameter m in any way. For example, a simple algorithm for sorting a list of n numbers would be in $\mathcal{O}(n^2)$, although there are faster algorithms that are in $\mathcal{O}(n \log n)$[1] This definition naturally extends to multiple problem parameters; for example, in Chapter 3 we discuss algorithms that run in time polynomial in the number of players, and in the maximal amount of resources players have.

---

[1]Note that this setting takes the number of numbers as the parameter, not the input size. The number of bits that are required to represent a number is not large: if numbers are written in binary, at most k bits are required to represent numbers of size $< 2^k$. Standard analysis assumes that comparing any two numbers takes a constant amount of time, and ignores input size (see Cormen et al. [2001] for details).

The class of problems that can be solved in time polynomial in the input size is referred to as P.

NP-complete problems [Garey and Johnson, 1979], are a famous class of computational problems. They are normally formulated as *decision problems*; that is, we are given an instance of a problem, and we must decide whether the answer to this problem is "yes" or "no". For example, the *vertex cover* problem [Garey and Johnson, 1979] is stated as follows: given a graph $\Gamma = \langle V, E \rangle$ and an integer k, is there a subset of vertices $C \subseteq V$ of size k such that for every vertex $v \in V$, it is either the case that $v \in C$ or v is incident to some $v' \in C$?

The class of NP-complete problems is interesting for several reasons. First, it is quite comprehensive, spanning the entire gamut of theoretical computer science. Second, they all have the "polynomial witness" property: it is possible to quickly verify whether a possible solution is correct. In the vertex cover problem above, it is very easy to guess a subset of vertices of size k, and check whether it is a vertex cover. Finally, it is currently not known whether there exist polynomial time algorithms that are able to solve NP-complete problems; however, it is generally believed that there are such algorithms exist, i.e., $P \neq NP$.

A problem X is NP-hard if there exists a *poly-time reduction* from an NP-complete problem Y to X. In formal terms, there exists some function $f : Y \to X$ such that for every $y \in Y$, $f(y)$ is a "yes" instance of X if and only if y is a "yes" instance of Y. We require that $\|f(y)\|$, is polynomial in $\|y\|$, and that the time it takes to compute $f(y)$ is also polynomial in y. Showing that a problem is NP-hard is a strong indication that no algorithm will decide it in polynomial time.

Chapter 2

Arbitration Functions and Stability in OCF Games

"I'm gonna make him an offer he can't refuse"

—Don Vito Corleone, *The Godfather*

In this chapter, we propose a formal framework for handling deviation in OCF games. Chalkiadakis et al. [2010] describe three ways in which non-deviators may react to deviation; we generalize these reactions using the *arbitration function*, a function that describes the amount each coalition (that contains non-deviators) assigns to deviators, given their deviation. We show that the three core concepts described by Chalkiadakis et al. [2010] are special cases of our model, and proceed to describe payoffs that are in cores of arbitrated OCF games. We then characterize core outcomes under some arbitration functions. First, we show that the conservative core (the core under the conservative arbitration function) is not empty if and only if an equivalent non-OCF game has a non-empty core. Second, using LP duality, we provide a necessary and sufficient condition for the sensitive and refined cores to be non-empty that is similar in spirit to the Bondareva-Shapley theorem [Bondareva, 1963, Shapley, 1967]. Using the characterization of the refined core, we identify a class of games that are guaranteed to have a non-empty refined core.

2.1 The Arbitration Function

Before diving into the theoretical underpinnings of arbitration functions, let us recall the motivation for their use. Consider a simple market exchange with three players. Player 1 can supply sugar to a bakery (owned by player 2), to a lemonade stand (owned by player 3), or to a coffee stall (owned by player 4). Player 1 has 100kg of sugar. All players need 50kg of sugar, but generate different revenue. Players 2,3,4 will generate $200, $300 and $400, respectively, by using the sugar. Now, suppose that player 1 chooses to provide sugar to players 3 and 4, selling it at $3/kg, i.e. player 1 receives $150 from each transaction. Player 1 can drop one of the transactions and sell 50kg of sugar for $3.5/kg to player 2 (or any price lower than $4/kg, and higher than $3/kg), and receive a strictly higher payoff. However, deciding whether doing so will result in a strictly higher payoff depends on how other players behave. For example, suppose that breaking the agreement with player 3 results in player 4 refusing to collaborate with player 1 as well; in that case, the most that player 1 can make by working with player 2 is $200, as opposed to the $300 he is getting now. Alternatively, it may be that the price of sugar is controlled by an external body, and any overcharging results in a heavy fine. To conclude, in settings where players devote partial resources to collaborative projects, non-deviators can affect the profitability of deviation.

In classic cooperative game theory, a coalition structure is simply a partition of N. A set S that is unhappy with its payoffs and wants to deviate compares its current payoff with its coalition value. However, as Example 1.6 shows, the complex structure of deviation in OCF games can lead to a variety of ways in which players may react to a deviation. If a deviating set still keeps some resources invested in coalitions with non-deviators, it may be the case that it will still be paid from those coalitions.

To discuss stability in OCF games, we need to describe how players react if some $S \subseteq N$ deviates from $(CS, \mathbf{x})$. One can think of the process in the following manner: given an outcome $(CS, \mathbf{x})$ and a deviation CS' of $S \subseteq N$ from CS, the set S may use all available resources in $CS|_S$ and CS' in order to generate revenue for itself. Moreover, each coalition $\mathbf{c}$ in $CS \setminus CS|_S$ needs to decide how much payoff it gives to the deviators; this behavior is specified by the *arbitration function*.

Definition 2.1. Given an outcome $(CS, \mathbf{x})$, a set $S \subseteq N$ and a deviation CS' of S from CS, the arbitration function $\mathcal{A}$ assigns a real value $\alpha_{\mathbf{c}}(CS, \mathbf{x}, S, CS')$ for each $\mathbf{c} \in CS \setminus CS|_S$.

The value $\alpha_{\mathbf{c}}(CS, \mathbf{x}, S, CS')$ represents the amount that the coalition $\mathbf{c}$ will pay S, given the current outcome $(CS, \mathbf{x})$, the identity of the deviators, and the nature of their deviation.

2.1.1 Properties of Arbitration Functions

We would naturally like to impose some restrictions on the amount that deviators may expect to get from non-deviators. For example, it would be unreasonable for a deviating set to expect a payoff of more than what the value of the coalition is, post deviation. Another natural restriction is that every non-deviator should receive the same payoff it did before the deviation, i.e., non-deviators will never agree to pay the deviating set S at the expense of any of the non-deviating members. Formally, we describe below some limitations on arbitration functions; given an outcome $(CS, \mathbf{x})$, a set $S \subseteq N$, a deviation CS' of S from CS and an arbitration function $\mathcal{A}$ whose output for the given outcomes is described by the list of functions $(\alpha_{\mathbf{c}})_{\mathbf{c} \in CS \setminus CS|_S}$, we require that $\mathcal{A}$ satisfies:

Accountability: $\alpha_{\mathbf{c}}(CS, \mathbf{x}, S, CS') \leq \max\{v(\mathbf{c} - \mathbf{d}(\mathbf{c})) - \sum_{i \in N \setminus S} x_i(\mathbf{c}), 0\}$

Deviation-Monotonicity: Given two subsets of $S \subseteq T \subseteq N$ and two deviations, CS' and CS'' from CS of S and T respectively such that for each $\mathbf{c} \in CS \setminus CS|_S$ we have that $\mathbf{d}_{CS'}(\mathbf{c}) \leq \mathbf{d}_{CS''}(\mathbf{c})$, then $\alpha_{\mathbf{c}}(CS, \mathbf{x}, S, CS') \geq \alpha_{\mathbf{c}}(CS, \mathbf{x}, T, CS'')$.

The first condition states a general upper bound on the amount that a deviating set can expect to receive from a coalition: after the deviation, the coalition $\mathbf{c}$ can generate a profit of $v(\mathbf{c}-\mathbf{d}(\mathbf{c}))$; the most that a deviating set can expect to receive is $v(\mathbf{c}-\mathbf{d}(\mathbf{c}))$, minus the original payments given to non-deviators under $(CS, \mathbf{x})$. The last condition simply states that an arbitration function cannot punish deviators more for hurting an agreement less: if the deviators withdraw less resources from each coalition, they should receive at least as much payoff as they would have received had their impact been greater. The rationale behind the first condition is strategic in nature; it stems from the assumption that a set of players engaged in a task will not agree to pay deviating members if the deviators cannot ensure that each non-deviator is paid the same amount he got before. We do stress however, that accountability is not a necessary component in our proofs: our results still hold even if one does not assume accountability. The second condition, while sensible as well, is instrumental in proving Theorem 2.3, and is a necessary assumption. Finally, note that it is possible that a coalition imposes fines on deviating members; that is, the value of $\alpha_{\mathbf{c}}$ need not be positive.

The total profit to a deviating set S, given the deviation CS' from CS is denoted

$$\mathcal{A}(CS, \mathbf{x}, S, CS') = v^*(\mathbf{w}(CS|_S) + \mathbf{w}(CS')) + \sum_{\mathbf{c} \in CS \setminus CS|_S} \alpha_{\mathbf{c}}(CS, \mathbf{x}, S, CS).$$

Given an outcome $(CS, \mathbf{x})$ and a set $S \subseteq N$, the most that S can get by deviating from $(CS, \mathbf{x})$ is denoted by

$$\mathcal{A}^*(CS, \mathbf{x}, S) = \sup\{\mathcal{A}(CS, \mathbf{x}, S, CS') \mid CS' \text{ is a deviation of } S \text{ from } CS\}.$$

Much like v^*, throughout this thesis we assume that $\mathcal{A}^*$ is achievable by some deviation of S from $(CS, \mathbf{x})$. This assumption holds if $\alpha_{\mathbf{c}}$ is a continuous (or bounded and piecewise continuous) function, since the space of all possible deviations of S from CS is compact. Now, in order for S to deem a deviation profitable, every one of its members must stand to gain from it. Formally, given an outcome $(CS, \mathbf{x})$ and a deviation CS' of $S \subseteq N$ from CS, CS' is *an $\mathcal{A}$-profitable deviation* if:

1. there exists a coalition structure CS_d such that $\mathbf{w}(CS_d) = \mathbf{w}(CS|_S) + \mathbf{w}(CS')$ and an imputation $\mathbf{x}_d \in I(CS_d)$.

2. there is a list of vectors $\mathbf{y} = \mathbf{y}_1, \ldots, \mathbf{y}_k \in \mathbb{R}^n$ such that for every $\mathbf{c} \in CS \setminus CS|_S$ we have that $supp(\mathbf{y}_\ell) \subseteq S \cap supp(\mathbf{c} - \mathbf{d}(\mathbf{c}))$. Moreover, for all $\mathbf{c} \in CS \setminus CS|_S$, $\sum_{i=1}^n y_i(\mathbf{c}) = \alpha_{\mathbf{c}}(CS, \mathbf{x}, S, CS')$; if $\alpha_{\mathbf{c}}(CS, \mathbf{x}, S, CS') \geq 0$ then $\mathbf{y}(\mathbf{c}) \in \mathbb{R}^n_+$ and it is in $\mathbb{R}^n_-$ otherwise.

3. for every $i \in S$ we have that $p_i(CS_d, \mathbf{x}_d) + \sum_{\mathbf{c} \in CS \setminus CS|_S} y_i(\mathbf{c}) > p_i(CS, \mathbf{x})$.

Simply put, given S's deviation from $(CS, \mathbf{x})$ and the arbitration function $\mathcal{A}$, the players in S must agree on three things: first, they must decide on a coalition structure to form with their post-deviation resources; second, they must agree on a way of dividing profits from that coalition structure and third, they must agree on a way of dividing the payoffs (or fines) from the arbitration function, such that all players in S receive strictly more than what they had received under $(CS, \mathbf{x})$. If no such payoff division exists, S cannot $\mathcal{A}$-profitably deviate.

The fact that $supp(\mathbf{y}(\mathbf{c})) \subseteq S \cap supp(\mathbf{c} - \mathbf{d}(\mathbf{c}))$ implies another property of arbitration functions: only those members of S who have their resources invested in $\mathbf{c}$ after S deviates can receive payoffs (or suffer fines) that originate from $\mathbf{c}$. Thus, if S withdraws all of its resources from $\mathbf{c}$, i.e., $\mathbf{d}(\mathbf{c}) = \mathbf{c}|_S$, then its payoff from $\mathbf{c}$ is 0. This assumption is in line with the coalitional efficiency assumption made on imputations in $I(CS)$. If we assume that $\alpha_\mathbf{c}$ can be shared among all members of S, then the model collapses to the setting where there is a single freely divisible payoff that S needs to share among its members (see Remark 2.2). We can make a slightly more relaxed assumption, which is that the members of $S \cap supp(\mathbf{c})$ share the value $\alpha_\mathbf{c}$ among themselves. This assumption is just as valid as the one we make and most of our results can be modified to suit this.

We do mention that one of the appealing points of having members of $\mathbf{c} - \mathbf{d}(\mathbf{c})$ share profits is that there exists an outcome that describes "the state of the world" post deviation. In slightly more formal terms, if the arbitration function $\mathcal{A}$ satisfies accountability, then for any deviation of a set S from an outcome $(CS, \mathbf{x})$, there exists an outcome where every player in S receives at least what he receives under the deviation (if $\mathcal{A}$ is non-negative, then each member of S receives exactly what he does under the deviation), and each member of $N \setminus S$ receives the same payoffs that he does under $(CS, \mathbf{x})$ from all coalitions in $CS|_{N \setminus S}$. This fact is used in Chapter 4, Lemma 4.11.

Remark 2.2. It is possible to incorporate a *freely divisible value* into the arbitration function. This value is a payoff (or fine) given to deviators by an external entity. Similar to the payments from coalitions, this payment may depend on the original outcome, the identity of the deviating set, and the nature of its deviation. All theorems in this paper carry through with the addition of a freely divisible value. Intuitively, this value extends the notion of such solution concepts as the ε-core [Maschler et al., 1979] to OCF games.

2.1.2 Some Types of Arbitration Functions

We now present some arbitration functions and briefly discuss their properties. The description of the conservative, refined and optimistic arbitration functions is first given by Chalkiadakis et al. [2010]; however, they do not use the term arbitration functions to describe them; the term, along with the general context for the conservative, refined and optimistic arbitration functions, is found in [Zick and Elkind, 2011].

The Conservative Arbitration Function The simplest assumption that one can make with respect to reaction to deviation is that non-deviators will react by voiding any agreement with deviators; that is, deviators may not expect any payment from any coalitions, regardless of their contribution. This reasoning gives rise to the *conservative arbitration function*, introduced by Chalkiadakis et al. [2010]; formally, $\alpha_{\mathbf{c}} \equiv 0$ for any given input. Under the conservative arbitration function, a deviating set is not rewarded in any way for continued interaction with non-deviators; thus, the best that it can hope to gain by deviating is the most that it can make on its own. Therefore, under the conservative arbitration function, denoted by $\mathcal{A}_c$, we have that $\mathcal{A}_c^*(CS, \mathbf{x}, S) = v^*(\mathbf{e}^S)$, for any outcome $(CS, \mathbf{x})$.

We stress that $\alpha_{\mathbf{c}} = 0$ even if S does not hurt any coalition, i.e., if $\mathbf{d}(\mathbf{c}) = 0^n$ for all $\mathbf{c} \in CS \setminus CS|_S$. In less formal terms: the mere mention of deviation under the conservative arbitration function is punished by complete banishment.

The Sensitive Arbitration Function When reasoning about player reaction to deviation, one can make the natural assumption that non-deviators would not mind what the deviating set S does, as long as the deviation does not affect them. Under this notion, called the *sensitive arbitration function*, the payment from the coalition $\mathbf{c} \in CS \setminus CS|_S$ is 0 if there is some $\mathbf{c}'$ such that $\mathbf{d}(\mathbf{c}') \neq 0$ and $supp(\mathbf{c}) \cap supp(\mathbf{c}') \cap (N \setminus S) \neq \emptyset$; otherwise, the payment to S is the original payoff that S receives from $\mathbf{c}$ under $(CS, \mathbf{x})$.

Under the sensitive arbitration function, denoted by $\mathcal{A}_s$, the most that a deviating set S can get can be computed as follows: first, it is clear that there is no benefit to S in investing resources in players that they hurt in some coalitions. Thus, S needs only to decide which players will it keep collaborating with and which players will it break all agreements with. In order to describe payments under the sensitive arbitration function, it is useful to use the following notation: given a set $T \subseteq N$, we let CS_T be the set of all coalitions in CS that involve members of T; that is

$$CS_T = (\mathbf{c} \in CS \mid supp(\mathbf{c}) \cap T \neq \emptyset).$$

Now, given a set $S \subseteq N$, if S chooses to break agreements that involve members of T, it foregoes all payments from coalitions in CS_T, but can use the resources it has invested in CS_T in order to maximize its own payments. In other words,

$$\mathcal{A}_s^*(CS, \mathbf{x}, S) = \max\{v^*(\mathbf{w}(CS|_S) + \mathbf{w}_S(CS_T)) + p_S(CS \setminus CS|_S \cup CS_T, \mathbf{x}) \mid T \subseteq N \setminus S\}.$$

The Refined Arbitration Function An even more lenient reaction to deviation is possible in the following scenario: a coalition $\mathbf{c}$ allows a deviating set to keep its payoffs from $\mathbf{c}$ if and only if S has not withdrawn any resources from it. This notion of deviation, termed the *refined arbitration function*, is more generous to the deviators than the sensitive arbitration function. Formally, given a deviation CS' of $S \subseteq N$ from CS in the outcome $(CS, \mathbf{x})$, the refined arbitration function, denoted $\mathcal{A}_r$, will let S keep its payoff from $\mathbf{c} \in CS \setminus CS|_S$ under $(CS, \mathbf{x})$ if $\mathbf{d}(\mathbf{c}) = 0^n$, and will pay it 0 otherwise.

The most that a set S can expect to gain by deviating under the refined arbitration function can be written explicitly as follows: first, given a coalition $\mathbf{c}$ in $CS \setminus CS|_S$, S should either withdraw all resources from $\mathbf{c}$ or none at all. Thus,

S needs to find the best set of coalitions in CS that contain $CS|_S$ from which to fully withdraw resources. Formally:

$$\mathcal{A}_r^*(CS, \mathbf{x}, S) = \max\{v^*(\mathbf{w}_S(\widehat{CS})) + p_S(CS \setminus \widehat{CS}, \mathbf{x}) \mid CS|_S \subseteq \widehat{CS} \subseteq CS\}.$$

The Optimistic Arbitration Function An even more lenient reaction to deviation is the *optimistic arbitration function*. Under the optimistic arbitration function, denoted $\mathcal{A}_o$, a coalition is willing to pay the deviating set, so long as non-deviators receive their original payoffs. Since every arbitration function $\mathcal{A}$ has the accountability property, this means that the optimistic arbitration function is giving a deviating set the highest possible payoff it can receive under any arbitration function. More specifically, given an outcome $(CS, \mathbf{x})$ and a deviation CS' of S from CS, we have that $\alpha_{\mathbf{c}}(CS, \mathbf{x}, S, CS') = \max\{v(\mathbf{c} - \mathbf{d}(\mathbf{c})) - \sum_{i \notin S} x_i(\mathbf{c}), 0\}$.

In order to compute the most that a set can get by deviating under the optimistic arbitration function, S needs to decide which coalitions it is going to withdraw all resources from, and which coalitions it is going to withdraw partial resources from. Formally:

$$\mathcal{A}_o^*(CS, \mathbf{x}, S) = \sup\{v^*(\mathbf{w}_S(\widehat{CS}) + \mathbf{w}(CS')) + \sum_{\mathbf{c} \in CS \setminus \widehat{CS}} v(\mathbf{c} - \mathbf{d}(\mathbf{c})) - p_{N \setminus S}(CS \setminus \widehat{CS}, \mathbf{x})\},$$

where $\widehat{CS}$ is a coalition structure containing $CS|_S$ (and possibly other coalitions of CS from which the members of S withdrew completely their resources), and CS' is a deviation of S from $CS \setminus \widehat{CS}$.

2.2 Redefining $\mathcal{A}$-Profitable Deviations

The definition of an $\mathcal{A}$-profitable deviation is somewhat difficult to work with; Theorem 2.3 provides us with a simpler, more intuitive notion of a profitable deviation. The main appeal of Theorem 2.3 is that it allows us to work with the values of $\mathcal{A}^*$ directly when assessing the stability of outcomes, rather than having to find an explicit division of payoffs that satisfy all players.

Theorem 2.3. *Given an OCF game $\mathcal{G} = \langle N, v \rangle$, an outcome $(CS, \mathbf{x})$ and an arbitration function $\mathcal{A}$, if there is a set $S \subseteq N$ such that $\mathcal{A}^*(CS, \mathbf{x}, S) > p_S(CS, \mathbf{x})$, then there is a subset of S that can $\mathcal{A}$-profitably deviate from $(CS, \mathbf{x})$.*

Proof. Suppose that there is a set $S \subseteq N$ such that $\mathcal{A}^*(CS, \mathbf{x}, S) > p_S(CS, \mathbf{x})$.

Let CS' be a deviation of S from $(CS, \mathbf{x})$ such that $\mathcal{A}^*(CS, \mathbf{x}, S) = \mathcal{A}(CS, \mathbf{x}, S, CS')$. We also let CS_d be a coalition structure such that $v(CS_d) = v^*(\mathbf{w}(CS|_S) + \mathbf{w}(CS'))$; both CS' and CS_d exist by our assumptions on $\mathcal{A}^*$ and v^*. Finally, let $\alpha_{\mathbf{c}}$ be the amount given to $S \cap supp(\mathbf{c} - \mathbf{d}(\mathbf{c}))$ after the deviation. Our objective is to find an imputation $\mathbf{x}_d \in I(CS_d)$ along with payoff divisions of $(\alpha_{\mathbf{c}})_{\mathbf{c} \in CS \setminus CS|_S}$, that will constitute an $\mathcal{A}$-profitable deviation of some subset of S.

For $\mathbf{c} \in CS \setminus CS|_S$, let $I_\mathbf{c}$ be the set of all possible ways that $\alpha_\mathbf{c}$ can be divided among the members of $S \cap supp(\mathbf{c} - \mathbf{d(c)})$, formally, if $\alpha_\mathbf{c}$ is positive, then:

$$I_\mathbf{c} = \{\mathbf{y(c)} \in \mathbb{R}_+^n \mid supp(\mathbf{y(c)}) \subseteq S \cap supp(\mathbf{c} - \mathbf{d(c)}); \sum_{i=1}^n y_i(\mathbf{c}) = \alpha_\mathbf{c}\};$$

If $\alpha_\mathbf{c}$ is negative then $I_\mathbf{c}$ is similarly defined, with all $\mathbf{y(c)} \in I_\mathbf{c}$ being vectors in $\mathbb{R}_-^n$. Given $\mathbf{x}_d \in I(CS_d)$ and $\mathbf{y} \in \prod_{\mathbf{c} \in CS \setminus CS|_S} I_\mathbf{c}$, let us write $q_i(\mathbf{x}_d, \mathbf{y})$ to be the payoff of i from the deviation, assuming payoffs are divided as per $\mathbf{x}_d$ and $\mathbf{y}$. We now define a function $TL : I(CS_d) \times \prod_{\mathbf{c} \in CS \setminus CS|_S} I_\mathbf{c} \to \mathbb{R}$ as follows:

$$TL(\mathbf{x}_d, \mathbf{y}) = \sum_{i \in S \mid p_i(CS, \mathbf{x}) > q_i(\mathbf{x}_d, \mathbf{y})} p_i(CS, \mathbf{x}) - q_i(\mathbf{x}_d, \mathbf{y}).$$

The function TL measures the total loss experienced by members of S who are getting less than what they are getting under $(CS, \mathbf{x})$; these are members of S that despite joining the deviators, do not enjoy a profit from deviating. Observe that TL is continuous, and that both $I(CS_d)$ and $\prod_{\mathbf{c} \in CS \setminus CS|_S} I_\mathbf{c}$ are compact spaces; thus, TL has a minimum value attained within its domain. Let us choose a minimizing point $(\bar{\mathbf{x}}, \bar{\mathbf{y}})$ where $\bar{\mathbf{x}} \in I(CS_d)$ and $\bar{\mathbf{y}} \in \prod_{\mathbf{c} \in CS \setminus CS|_S} I_\mathbf{c}$, for which the number of players who strictly benefit from deviating is maximal. That is, $(\bar{\mathbf{x}}, \bar{\mathbf{y}})$ are both a global minimum of TL, and they achieve this global minimum in a manner that minimizes the number of players in S that suffer a loss due to their deviation.

Using $\bar{\mathbf{x}}$ and $\bar{\mathbf{y}}$, we argue that there exists a subset of S that can $\mathcal{A}$-profitably deviate from $(CS, \mathbf{x})$. Let us construct a directed graph Γ in the following manner: the vertices of Γ are the members of S, and there is an edge from i to i' if i can legally transfer payoff to i' under $\bar{\mathbf{x}}$ and $\bar{\mathbf{y}}$. This happens if both i and i' are in the support of the same coalition (either a coalition $\mathbf{c}$ in CS_d or some coalition $\mathbf{c} - \mathbf{d(c)}$, for some $\mathbf{c} \in CS \setminus CS|_S$), and i receives a positive payoff from that coalition. Alternatively, if a coalition $\mathbf{c} \in CS \setminus CS|_S$ assigns a negative payoff to S (the value $\alpha_\mathbf{c}$ is negative), then there is an edge from i to i' if both are in the support of $\mathbf{c}$ and the payoff of i' from that coalition is negative.

We say that a vertex (player) $i \in S$ is green if $p_i(CS, \mathbf{x}) < q_i(\bar{\mathbf{x}}, \bar{\mathbf{y}})$; red if $p_i(CS, \mathbf{x}) > q_i(\bar{\mathbf{x}}, \bar{\mathbf{y}})$ and white if equality holds. Green vertices strictly benefit from the proposed payoff division, red vertices strictly lose, while white vertices break even.

We make the following critical observation: consider a vertex $i \in S$; if there is an edge (i, i') in Γ, then there is some $\varepsilon > 0$ that i can transfer to i', such that the resulting payoff division is valid. This immediately implies that for any green vertex g and red vertex r, there is no edge (g, r); otherwise, g could have transferred a small positive payoff to r while remaining green, and the resulting payoff division will have a strictly lower value for TL than $(\bar{\mathbf{x}}, \bar{\mathbf{y}})$. Moreover, for any green vertex g and any white vertex w, if (g, w) is an edge, then g can transfer a small amount to w, making him green as well, while keeping the same value for the function TL; this contradicts the fact that the number of green players is maximal under $(\bar{\mathbf{x}}, \bar{\mathbf{y}})$. We conclude that if a vertex i is not green, then there is no directed edge from any green vertex g to i. If that is the case, then if both g and i are in the support of some coalition after the deviation, then g receives no payoff from that coalition. A similar argument shows that if $\alpha_{\mathbf{c}}$ is negative and g and i are in the support of $\mathbf{c}$ then i receives no negative payoff from that coalition; that is, if a coalition assigns a negative payoff to deviating players, then green players are the only players that incur a loss from that coalition.

Consider the set of green vertices in S, denoted S_g. Observe that the players in S_g receive no payoff from coalitions they share with non-green players in S. We claim that S_g can $\mathcal{A}$-profitably deviate from $(CS, \mathbf{x})$. Indeed, suppose that the players in $S \setminus S_g$ do not deviate. Suppose also that the resources that S_g withdrew from CS in the deviation CS' in order to invest in coalitions with members of $S \setminus S_g$ in CS_d are kept as is. Since $\mathcal{A}$ is deviation-monotone, the payment to S_g from the coalitions it alone deviated from is now weakly higher. Thus, if S_g behaves in the same way it did under the deviation CS', it forms the same coalitions as in $CS_d|_{S_g}$, and it divides payoffs as per $(\bar{\mathbf{x}}, \bar{\mathbf{y}})$, then all of its members are strictly better off. Finally, $S_g \neq \emptyset$, since $p_S(CS, \mathbf{x}) < \mathcal{A}^*(CS, \mathbf{x}, S)$, which concludes the proof. □

Simply put, Theorem 2.3 states that if the most that a set S can get by deviating from $(CS, \mathbf{x})$ is strictly greater than its payoff from $(CS, \mathbf{x})$, then there is some non-empty subset of S which can $\mathcal{A}$-profitably deviate. Note that the distinction between coalitional value and the profitability of deviating exists in cooperative games with coalition structures as well.

Example 2.4. Consider a modified version of the three player 2-majority game: the player set is $N = \{1, 2, 3\}$, the value of any coalition $S \subseteq N$ of size at least 2 is 1, and the value of singletons is $0 < \varepsilon < \frac{1}{3}$. Suppose that players form the coalition N, and divide payoffs such that each player receives $\frac{1}{3}$. The coalition N can gain more by forming the coalition structure $(\{1, 2\}, \{3\})$; however, there is no way to divide payoffs such that every player receives more than $\frac{1}{3}$: player 3 can receive only ε, hence he cannot be better off.

While the set N cannot profitably deviate, it contains a strict subset (namely, any subset of N whose size is 2) that can profitably deviate.

Theorem 2.3 provides us with a good justification for using $\mathcal{A}^*(CS, \mathbf{x}, S)$ as a measure of the satisfaction of S with an outcome. If S can $\mathcal{A}$-profitably deviate from $(CS, \mathbf{x})$ via some deviation CS', then clearly $\mathcal{A}^*(CS, \mathbf{x}, S) > p_S(CS, \mathbf{x})$. On the other hand, if $\mathcal{A}^*(CS, \mathbf{x}, S) > p_S(CS, \mathbf{x})$, then there is some $T \subseteq S$ that can $\mathcal{A}$-profitably deviate from $(CS, \mathbf{x})$.

2.3 The Arbitrated Core

Given an OCF game $\mathcal{G} = \langle N, v \rangle$ and an arbitration function $\mathcal{A}$, we say that an outcome $(CS, \mathbf{x})$ is $\mathcal{A}$-*stable* if there is no set $S \subseteq N$ that can $\mathcal{A}$-profitably deviate from $(CS, \mathbf{x})$. The *arbitrated core* of $\mathcal{G}$ w.r.t. some $\mathcal{A}$, or the $\mathcal{A}$-core of $\mathcal{G}$ (denoted

$Core(\mathcal{G}, \mathcal{A}))$, is the set of all $\mathcal{A}$-stable outcomes over $\mathcal{G}$. We replace $\mathcal{A}_c$-stable with c-stable when discussing stability in the conservative core, and similarly refer to stability under the sensitive, refined and optimistic arbitration functions as s, r and o-stable. This is in accordance with the notation used by Chalkiadakis et al. [2010].

Using Theorem 2.3, we can completely characterize $\mathcal{A}$-stable outcomes.

Theorem 2.5. *Given an OCF game $\mathcal{G} = \langle N, v \rangle$, an outcome $(CS, \mathbf{x})$ is in the $\mathcal{A}$-core of $\mathcal{G}$ if and only if for every $S \subseteq N$ we have $p_S(CS, \mathbf{x}) \geq \mathcal{A}^*(CS, \mathbf{x}, S)$.*

Proof. First, suppose that for every $S \subseteq N$ we have that $p_S(CS, \mathbf{x}) \geq \mathcal{A}^*(CS, \mathbf{x}, S)$. This means that for any deviation of S from CS, say CS', we have that $p_S(CS, \mathbf{x}) \geq \mathcal{A}(CS, \mathbf{x}, S, CS')$: the total payoff to S from deviating is no more than $\mathcal{A}^*(CS, \mathbf{x}, S)$, which is by assumption no more than $p_S(CS, \mathbf{x})$. This implies that no matter how S divides the deviation payoffs, there would be at least one $i \in S$ who gets no more than $p_i(CS, \mathbf{x})$, thus S cannot $\mathcal{A}$-profitably deviate.

On the other hand, suppose that there exists some $S \subseteq N$ such that $p_S(CS, \mathbf{x}) < \mathcal{A}^*(CS, \mathbf{x}, S)$; by Theorem 2.3, there is some subset of S that can $\mathcal{A}$-profitably deviate from $(CS, \mathbf{x})$, thus $(CS, \mathbf{x})$ is not $\mathcal{A}$-stable. □

Theorem 2.5 immediately implies the following claim: the cores induced by more lenient arbitration functions are contained in cores with stricter arbitration functions. Intuitively, this is obvious: if an outcome is stable w.r.t. $\mathcal{A}$, $\mathcal{A}$ always pays more to deviators than $\bar{\mathcal{A}}$, and $(CS, \mathbf{x})$ is $\mathcal{A}$-stable, then it is $\bar{\mathcal{A}}$-stable as well. Formally, given two arbitration functions $\mathcal{A}$ and $\bar{\mathcal{A}}$, if for any set $S \subseteq N$ and any outcome $(CS, \mathbf{x})$ we have that $\mathcal{A}^*(CS, \mathbf{x}, S) \geq \bar{\mathcal{A}}^*(CS, \mathbf{x}, S)$, then we say that $\mathcal{A} \geq \bar{\mathcal{A}}$. Using this definition, we obtain the following corollary:

Corollary 2.6. *If $\mathcal{A} \geq \bar{\mathcal{A}}$ then $Core(\mathcal{G}, \mathcal{A}) \subseteq Core(\mathcal{G}, \bar{\mathcal{A}})$.*

Proof. Suppose that $(CS, \mathbf{x}) \in Core(\mathcal{G}, \mathcal{A})$; this means that for all $S \subseteq N$ we have that $p_S(CS, \mathbf{x}) \geq \mathcal{A}^*(CS, \mathbf{x}, S)$. Since $\mathcal{A} \geq \bar{\mathcal{A}}$, we must have that $p_S(CS, \mathbf{x}) \geq \bar{\mathcal{A}}^*(CS, \mathbf{x})$ for all $S \subseteq N$ as well, which implies that $(CS, \mathbf{x}) \in Core(\mathcal{G}, \bar{\mathcal{A}})$. □

The conservative, sensitive, refined and optimistic arbitration functions described in Section 2.1.2, denoted $\mathcal{A}_c, \mathcal{A}_s, \mathcal{A}_r$ and $\mathcal{A}_o$, respectively, satisfy $\mathcal{A}_o \geq \mathcal{A}_r \geq \mathcal{A}_s \geq \mathcal{A}_c$. Corollary 2.6 implies that the conservative core contains the sensitive core, which contains the refined core, which contains the optimistic core. Similar observations appear in [Chalkiadakis et al., 2010] as well; in fact, it is shown by Chalkiadakis et al. [2010] that this containment can be strict for the optimistic, refined and conservative cores, i.e., there are games for which the conservative core strictly contains the refined core, and there are games where the refined core strictly contains the optimistic core. We mention that a similar separation can be shown for the sensitive core.

Finally, if we observe the class of arbitration functions that have the accountability property (see Section 2.1.1), then the optimistic arbitration function is the most generous arbitration function possible, i.e., for any arbitration function $\mathcal{A}$ satisfying accountability, $\mathcal{A}_o \geq \mathcal{A}$. This immediately implies that if an outcome $(CS, \mathbf{x})$ is in the o-core, then it is in the $\mathcal{A}$-core for any $\mathcal{A}$ that satisfies accountability.

2.3.1 Some Arbitrated Cores

Using the characterization result in Theorem 2.5, we now proceed to describe some arbitrated cores, namely those corresponding to the arbitration functions described in Section 2.1.2.

The Conservative Core

As argued in Section 2.1.2, for any outcome $(CS, \mathbf{x})$ and any set $S \subseteq N$, the most that S can get under the conservative arbitration function is $v^*(\mathbf{e}^S)$. Thus, the conservative core is the set of all outcomes $(CS, \mathbf{x})$ such that for every set $S \subseteq N$ we have that

$$p_S(CS, \mathbf{x}) \geq v^*(\mathbf{e}^S). \tag{2.1}$$

The Sensitive Core

In order for an outcome to be stable under the sensitive arbitration function, it must be that for any set $S \subseteq N$ and any $T \subseteq N \setminus S$ we have that

$$p_S(CS, \mathbf{x}) \geq v^*(\mathbf{w}(CS|_S) + \mathbf{w}_S(CS_T)) + p_S(CS \setminus CS|_S \cup CS_T, \mathbf{x});$$

where CS_T is the set of all coalitions in CS that involve members of T. Subtracting payments from $CS \setminus CS|_S \cup CS_T$ from both sides of the inequality, we get

$$p_S(CS|_S, \mathbf{x}) + p_S(CS_T, \mathbf{x}) \geq v^*(\mathbf{w}_S(CS|_S) + \mathbf{w}_S(CS_T)).$$

Note that by efficiency, $p_S(CS|_S, \mathbf{x}) = v(CS|_S)$. Moreover, if CS is an optimal coalition structure, then it must be the case that $p_S(CS|_S, \mathbf{x}) = v^*(CS|_S)$. We note that if CS is not an optimal coalition structure then $(CS, \mathbf{x})$ is not in the conservative core (as shown by Chalkiadakis et al. [2010]), and in particular is not stable with respect to the sensitive arbitration function. Therefore, we can assume that CS is an optimal coalition structure, and thus that $CS|_S$ is optimal for any $S \subseteq N$. Using this, we get that in order for an outcome to be in the sensitive core, it must hold that for any $S \subseteq N$ and any $T \subseteq N \setminus S$,

$$p_S(CS_T, \mathbf{x}) \geq v^*(\mathbf{w}_S(CS|_S) + \mathbf{w}_S(CS_T)) - v^*(\mathbf{w}_S(CS|_S)), \tag{2.2}$$

i.e. the total payoff to S from CS_T must be at least the marginal benefit of using the resources S has invested in CS_T. Note that if we take $T = N \setminus S$ then we obtain the conservative core condition.

The Refined Core

For the refined core, we have that

$$\mathcal{A}_r^*(CS, \mathbf{x}, S) = \max\{v^*(\mathbf{w}_S(\widehat{CS})) + p_S(CS \setminus \widehat{CS}, \mathbf{x}) \mid CS|_S \subseteq \widehat{CS} \subseteq CS\}.$$

Thus, an outcome $(CS, \mathbf{x})$ is in the refined core if and only if for any $S \subseteq N$ and for any coalition structure $\widehat{CS}$ containing $CS|_S$ we have that

$$p_S(CS, \mathbf{x}) \geq v^*(\mathbf{w}_S(\widehat{CS})) + p_S(CS \setminus \widehat{CS}, \mathbf{x}).$$

Arbitration, Fairness and Stability

This immediately implies that $(CS, \mathbf{x})$ is in the refined core if and only if for any such S and $\widehat{CS}$ we have

$$p_S(\widehat{CS}, \mathbf{x}) \geq v^*(\mathbf{w}_S(\widehat{CS})). \tag{2.3}$$

In other words, the payoff to S from any coalition structure $\widehat{CS}$ containing $CS|_S$ must be more than the most that S can get from the resources it has invested in $\widehat{CS}$. Limiting ourselves to coalition structures $\widehat{CS}$ such that $\widehat{CS} = CS|_S \cup CS_T$ for some $T \subseteq N \setminus S$ gives us exactly the sensitive core condition.

The Optimistic Core

For the optimistic arbitration function, $\mathcal{A}_o$ we have shown that

$$\mathcal{A}_o^*(CS, \mathbf{x}, S) = \sup\{v^*(\mathbf{w}_S(\widehat{CS}) + \mathbf{w}(CS')) + \sum_{\mathbf{c} \in CS \setminus \widehat{CS}} v(\mathbf{c} - \mathbf{d}(\mathbf{c})) - p_{N \setminus S}(CS \setminus \widehat{CS}, \mathbf{x})\},$$

where $\widehat{CS}$ is a coalition structure containing $CS|_S$, and CS' is a deviation of S from $CS \setminus \widehat{CS}$. In order for an outcome $(CS, \mathbf{x})$ to be in the optimistic core, it must be that for any set $S \subseteq N$, any $\widehat{CS} \subseteq CS$ containing $CS|_S$ and any CS' that is a deviation of S from $CS \setminus \widehat{CS}$ we have that

$$p_S(CS, \mathbf{x}) \geq v^*(\mathbf{w}_S(\widehat{CS}) + \mathbf{w}(CS')) + \sum_{\mathbf{c} \in CS \setminus \widehat{CS}} v(\mathbf{c} - \mathbf{d}(\mathbf{c})) - p_{N \setminus S}(CS \setminus \widehat{CS}, \mathbf{x}).$$

Let us write $CS \setminus \widehat{CS} - CS'$ to be the coalition structure $CS \setminus \widehat{CS}$ after S has withdrawn resources from it according to CS'; in that case, $\sum_{\mathbf{c} \in CS \setminus \widehat{CS}} v(\mathbf{c} - \mathbf{d}(\mathbf{c})) = v(CS \setminus \widehat{CS} - CS')$. Moreover, $p_N(CS \setminus \widehat{CS}, \mathbf{x}) = v(CS \setminus \widehat{CS})$. We conclude that the above equation is equivalent to

$$p_S(\widehat{CS}, \mathbf{x}) \geq v^*(\mathbf{w}_S(\widehat{CS}) + \mathbf{w}(CS')) + v(CS \setminus \widehat{CS} - CS') - v(CS \setminus \widehat{CS}). \tag{2.4}$$

The optimistic core stability condition is thus similar to the refined core stability condition; however, S is also allowed to withdraw resources from coalitions outside of $\widehat{CS}$ if it is willing to assume the marginal costs of this withdrawal.

2.4 Non-emptiness of Arbitrated Cores

Core outcomes are highly desirable both in classic cooperative games and in OCF games. Thus, it is important to find conditions on the characteristic function of an OCF game that would guarantee non-emptiness of its arbitrated core, as well as to identify classes of OCF games whose arbitrated core is always non-empty, for natural arbitration functions. We note that both of these questions can be — and have been— looked at through the lens of computational complexity. That is, one can ask whether there exist *polynomial-time* algorithms that decide whether the arbitrated core is non-empty or find an outcome in the arbitrated core, either for the general model or for specific classes of OCF games. We return to this question in Chapter 3.

Our characterization results are similar in spirit to the celebrated Bondareva–Shapley condition for core non-emptiness in classic cooperative games Bondareva [1963], Shapley [1967]. We will now briefly state this condition. Given a set $N = \{1, \ldots, n\}$, a collection of weights $(\delta_S)_{S \subseteq N}$ is called *balanced* if $\delta_S \geq 0$ for all $S \subseteq N$ and $\sum_{S: i \in S} \delta_S = 1$ for all $i \in N$. We can view a balanced collection of weights as a way for a player i to partially participate in all sets that contain i, with its contribution specified by the respective weight.

Given a cooperative game $\mathcal{G} = \langle N, u \rangle$ where $u : 2^N \to \mathbb{R}_+$, we say that $\mathcal{G}$ is *balanced* if for every balanced collection of weights $(\delta_S)_{S \subseteq N}$ it holds that $\sum_{S \subseteq N} \delta_S u(S) \leq u(N)$.

Theorem 2.7 (Bondareva–Shapley Theorem). *A cooperative game $\mathcal{G} = \langle N, u \rangle$ with $u : 2^N \to \mathbb{R}_+$ has a non-empty core if and only if it is balanced.*

An important feature of this characterization is that it is stated in terms of the characteristic function itself, and does not explicitly refer to outcomes of the game. In what follows, we describe similar characterizations for OCF games, for the conservative, sensitive and refined arbitration functions.[1] Specifically, for each of these arbitration functions, given an OCF game $\mathcal{G} = \langle N, v \rangle$ and a coalition structure $CS \in \mathcal{CS}(N)$, we characterize the set of imputations $\mathbf{x} \in I(CS)$ such that $(CS, \mathbf{x})$ is in the respective arbitrated core of $\mathcal{G}$. We limit our attention to optimal coalition structures, i.e., we assume that $v(CS) = v^*(\mathbf{e}^N)$: if CS is not optimal, then no outcome of the form $(CS, \mathbf{x})$ can be in the arbitrated core of $\mathcal{G}$ for any of our arbitration functions, since the grand coalition itself can profitably deviate from $(CS, \mathbf{x})$ [Chalkiadakis et al., 2010].

2.4.1 The Conservative Core

Given an OCF game $\mathcal{G} = \langle N, v \rangle$ and a coalition structure $CS = (\mathbf{c}_1, \ldots, \mathbf{c}_m)$, we say that a collection of non-negative weights $\{(r_j)_{j=1}^m; (\delta_S)_{S \subseteq N}\}$ is *c-balanced with respect to CS* if for every $i \in N$ and every coalition $\mathbf{c}_j$ such that $i \in supp(\mathbf{c}_j)$ it holds that $r_j + \sum_{S: i \in S} \delta_S = 1$. Chalkiadakis et al. [2010] show the following result.

Theorem 2.8 (Theorem 2, Chalkiadakis et al. [2010]). *Given an OCF game $\mathcal{G} = \langle N, v \rangle$ and an optimal coalition structure $CS \in \mathcal{CS}(N)$, there exists an outcome $\mathbf{x} \in I(CS)$ such that $(CS, \mathbf{x})$ is in the conservative core of $\mathcal{G}$ if and only if for every collection of non-negative weights $\{(r_j)_{j=1}^m; (\delta_S)_{S \subseteq N}\}$ that is c-balanced with respect to CS it holds that*

$$\sum_{j=1}^m r_j v(\mathbf{c}_j) + \sum_{S \subseteq N} \delta_S v^*(\mathbf{e}^S) \leq v^*(\mathbf{e}^N).$$

We begin by mentioning that Theorem 2.8 has a minor bug. The proof of Theorem 2.8 is as follows. Given an optimal coalition structure $CS = (\mathbf{c}_1, \ldots, \mathbf{c}_m)$,

---

[1] The case of the conservative arbitration function has been considered by Chalkiadakis et al. [2010]; we reproduce their results for completeness and to facilitate the comparison with the other two cases.

consider the following linear program.

$$\min \quad \sum_{j=1}^{m} \sum_{i \in supp(\mathbf{c}_j)} x_{ji} \qquad (2.5)$$

$$\text{s.t.} \quad \sum_{i \in supp(\mathbf{c}_j)}^{m} x_{ji} \geq v(\mathbf{c}_j) \quad \forall j \in \{1,\ldots,m\}$$

$$\sum_{i \in S} \sum_{j=1}^{m} x_{ji} \geq v^*(\mathbf{e}^S) \quad \forall S \subseteq N \qquad (2.6)$$

The dual of LP (2.6) is

$$\max \quad \sum_{j=1}^{m} r_j v(\mathbf{c}_j) + \sum_{S \subseteq N} \delta_S v^*(\mathbf{e}^S) \qquad (2.7)$$

$$\text{s.t.} \quad r_j + \sum_{S: i \in S} \delta_S = 1 \quad \forall j, \forall i \in supp(\mathbf{c}_j)$$

$$r_j \geq 0 \quad \forall j \in \{1,\ldots,m\}$$

$$\delta_S \geq 0 \quad \forall S \subseteq N \qquad (2.8)$$

LP (2.6) describes the constraints of the conservative core, with the exception that $\sum_{i \in supp(\mathbf{c}_j)} x_{ji} \geq v(\mathbf{c}_j)$, rather than an equality. Thus, if $\bar{\mathbf{x}}$ is an optimal solution to LP (2.6), then $(CS, \bar{\mathbf{x}})$ is in the c-core if and only if $\sum_{j=1}^{m} \sum_{i \in supp(\mathbf{c}_j)} x_{ji} = v(CS) = v^*(\mathbf{e}^N)$. By utilizing LP duality, we obtain that LP (2.7) has a value of at most $v^*(\mathbf{e}^N)$ if and only if CS can be stabilized w.r.t. the conservative arbitration function. The bug in Theorem 2.8 lies in the requirement that x_{ji} are unconstrained, which allows us to have $r_j + \sum_{S:i \in S} \delta_S = 1$ rather than $r_j + \sum_{S:i \in S} \delta_S \leq 1$. However, if we allow the variables x_{ji} to be unconstrained, and possibly strictly less than 0, we are effectively breaking the payoff distribution requirement for imputations. In the Bondareva-Shapley theorem, we could assume that the payoffs to players were unconstrained, since we have that $p_i \geq u(\{i\}) \geq 0$, which means that this assumption is implicitly made. However, for the conservative core we only get that $\sum_{j:i \in supp(\mathbf{c}_j)} x_{ji} \geq v^*(\mathbf{e}^{\{i\}})$ for all $i \in N$, which says nothing about the individual payments from coalitions.

We begin by providing a correction to this issue. The problem with Theorem 2 in [Chalkiadakis et al., 2010] is that it shows the existence of a c-stable *preimputation* for a c-balanced OCF game; formally, given a coalition structure $CS = (\mathbf{c}_1, \ldots, \mathbf{c}_m)$, a preimputation $\mathbf{x} = (\mathbf{x}_1, \ldots, \mathbf{x}_m)$ is a list of vectors in $\mathbb{R}^n$ that satisfies $\sum_{i=1}^{n} x_{ji} = v(\mathbf{c}_j)$, and if $c_{ji} = 0$ then $x_{ji} = 0$. That is, a preimputation satisfies the same conditions as an imputation, but drops the non-negativity requirement. The set of preimputations for a coalition structure CS is denoted $I_{pre}(CS)$.

Theorem 2.9. *If there exists a preimputation* $\mathbf{x}$ *that satisfies the constraints of LP (2.6), then there exists an imputation* $\bar{\mathbf{x}}$ *such that* $p_i(CS, \mathbf{x}) = p_i(CS, \bar{\mathbf{x}})$ *for all* $i \in N$.

Proof. Given a point $\mathbf{x} \in I_{pre}(CS)$, let us define

$$TN(\mathbf{x}) = \sum_{j=1}^{m} \sum_{i=1}^{n} \min\{0, x_{ji}\}.$$

As in Theorem 2.3, TN measures the total negative values of x_{ji}. Fixing an arbitrary point $\mathbf{x} \in I_{pre}(CS)$, let us observe the set

$$I_{pre}(CS, \mathbf{x}) = \{\mathbf{y} \in I_{pre}(CS), p_i(CS, \mathbf{y}) = p_i(CS, \mathbf{x}) \text{ for all } i \in N \mid TN(\mathbf{y}) \geq TN(\mathbf{x})\}.$$

$I_{pre}(CS, \mathbf{x})$ is compact and TN is continuous, so there exists some point $\bar{\mathbf{x}} \in I_{pre}(CS, \mathbf{x})$ that maximizes TN over $I_{pre}(CS, \mathbf{x})$.

Given a point $\mathbf{x} \in I_{pre}(CS)$, we define a graph $\Gamma(\mathbf{x}) = \langle V, E \rangle$ as follows: the vertices of $\Gamma(\mathbf{x})$ are
$$V = \{(j, i) \mid \mathbf{c}_j \in CS, i \in supp(\mathbf{c}_j)\},$$
and $E = E_1 \cup E_2$, where
$$E_1 = \{((i, j), (i, j')) \mid i \in supp(\mathbf{c}_j) \cap supp(\mathbf{c}_{j'}), j \neq j'\},$$
and
$$E_2 = \{((i, j), (i', j)) \mid i, i' \in supp(\mathbf{c}_j), x_{ji} > 0\}.$$

In simple terms, E_1 consists of all points $(i, j), (i, j')$ where i is in both $supp(\mathbf{c}_j)$ and $supp(\mathbf{c}_{j'})$, and there exists an edge from (i, j) to (i', j) in E_2 if $i, i' \in supp(\mathbf{c}_j)$ for some $\mathbf{c}_j \in CS$, and $x_{ji} > 0$. This means that i can transfer some small amount to i' under the coalition $\mathbf{c}_j$, and still maintain a positive payoff from $\mathbf{c}_j$.

Suppose that under $\bar{\mathbf{x}} \in \arg\max_{\mathbf{y} \in I_{pre}(CS, \mathbf{x})} TN(\mathbf{y})$, there is some coalition $\mathbf{c}_{j^*}$ and some $i^* \in supp(\mathbf{c}_{j^*})$ such that $\bar{x}_{j^*i^*} < 0$.

Let us observe a minimal length directed cycle in $\Gamma(\bar{\mathbf{x}})$ containing an edge $((i^*, j^*), (i^*, j_0))$, that contains edges from E_2. First, there must be outgoing edges from (i^*, j^*); otherwise, i^* is only being paid by $\mathbf{c}_{j^*}$. In particular, this would mean that i^* receives a strictly negative payoff under $(CS, \bar{\mathbf{x}})$, a contradiction to the fact that $p_i(CS, \bar{\mathbf{x}}) \geq v^*(\mathbf{e}^{\{i\}}) \geq 0$. Now, if a cycle C containing an outgoing edge from (i^*, j^*) and intersecting E_2 is of minimum length, it contains no path of the form
$$(i, j_1) \to (i, j_2) \to (i, j_3);$$
this is because the edge $((i, j_1), (i, j_3))$ is in E_1, a contradiction to C being of minimum length. Moreover, C cannot contain a path of the form
$$(i_1, j) \to (i_2, j) \to (i_3, j);$$
this is because if $((i_1, j), (i_2, j))$ is in E_2, then the player i_1 receives a positive payoff from the coalition $\mathbf{c}_j$, and in particular the edge $((i_1, j), (i_3, j))$ is also in E_2, a contradiction to C being the minimum length cycle intersecting with E_2.

We conclude that a minimum length cycle containing an outgoing edge from (i^*, j^*) and that intersects E_2 must be of the form
$$(i^*, j^*) \to (i^*, j_0) \to (i_1, j_0) \to (i_1, j_1) \to (i_2, j_1) \to \cdots \to (i_m, j^*) \to (i^*, j^*).$$

This means that i^* receives a positive payoff from $\mathbf{c}_{j_0}$, i_1 receives a positive payoff from $\mathbf{c}_{j_1}$, i_2 receives a positive payoff from $\mathbf{c}_{j_2}$ and, in general, i_ℓ receives a positive payoff from the coalition $\mathbf{c}_{j_\ell}$ for all $1 \leq \ell \leq m - 1$. Finally, the player i_m receives a positive payoff from $\mathbf{c}_{j^*}$.

Thus, i^* can transfer some small payoff of $\varepsilon > 0$ to i_1 under $\mathbf{c}_{j_0}$, i_1 can transfer ε to i_2 under $\mathbf{c}_{j_1}$ and so on, and i_m can transfer ε to i^* under $\mathbf{c}_{j^*}$. For any ε satisfying
$$0 < \varepsilon < \min\{\bar{x}_{j_0, i^*}, \bar{x}_{j_1, i_1}, \ldots, \bar{x}_{j_{m-1}, i_{m-1}}, \bar{x}_{j^*, i_m}\},$$

the resulting preimputation, $\mathbf{y}$, will have $p_i(CS, \mathbf{y}) = p_i(CS, \bar{\mathbf{x}})$ for all $i \in N$, and is therefore an optimal solution to LP (2.6). However, $TN(\mathbf{y}) > TN(\bar{\mathbf{x}})$, a contradiction to the fact that $\bar{\mathbf{x}}$ is a maximum of TN over $I_{pre}(CS, \mathbf{x})$. We conclude that $\Gamma(\bar{\mathbf{x}})$ contains no cycles that contain an outgoing edge from (i^*, j^*) and that intersect E_2.

Let us define
$$N_p = \{i \in N \mid \text{there is a path from } (j^*, i^*) \text{ to } (j, i) \text{ for some } \mathbf{c}_j\},$$

then for every $i \in N_p$, and every $i' \in N \setminus N_p$, there is no edge of the form $((i, j), (i', j))$. In other words, if $i, i' \in supp(\mathbf{c}_j)$ then $\bar{x}_{ji} \leq 0$. In particular, this means that $p_{N_p}(CS, \bar{\mathbf{x}}) \leq p_{N_p}(CS|_{N_p}, \bar{\mathbf{x}})$. Moreover, since $x_{j^*, i^*} < 0$, and $i^* \in N_p$, the above inequality is strict. Now,
$$p_{N_p}(CS|_{N_p}, \bar{\mathbf{x}}) = v(CS|_{N_p}) \leq v^*(\mathbf{e}^{N_p});$$

combining the inequalities we obtain that $p_{N_p}(CS, \bar{\mathbf{x}}) < v^*(\mathbf{e}^{N_p})$, which is a contradiction to $(CS, \bar{\mathbf{x}})$ satisfying the inequalities in LP (2.6).

We conclude that $\bar{\mathbf{x}}$ is an imputation of CS such that $p_i(CS, \mathbf{x}) = p_i(CS, \bar{\mathbf{x}})$ for all $i \in N$, which concludes that proof. $\square$

Example 2.10. Suppose the players form the coalition structure
$$CS = \left(\mathbf{c}_1 = \begin{pmatrix} 0.6 \\ 0.3 \\ 0.3 \end{pmatrix} \quad \mathbf{c}_2 = \begin{pmatrix} 0.4 \\ 0.5 \\ 0 \end{pmatrix} \quad \mathbf{c}_3 = \begin{pmatrix} 0.1 \\ 0.1 \\ 0.7 \end{pmatrix}\right)$$

and use the imputation
$$\mathbf{x} = \left(\mathbf{x}_1 = \begin{pmatrix} 0 \\ 2 \\ 3 \end{pmatrix} \quad \mathbf{x}_2 = \begin{pmatrix} 2 \\ 2 \\ 0 \end{pmatrix} \quad \mathbf{x}_3 = \begin{pmatrix} 3 \\ 0 \\ -1 \end{pmatrix}\right)$$

The resulting graph is shown in Figure 2.1, and the shortest cycle starting from $(3,3)$ including edges in E_2 is
$$(3,3) \to (3,1) \to (1,1) \to (1,3) \to (3,3).$$

That is, the negative payoff to player 3 from $\mathbf{c}_3$ can be reduced, if player 3 transfers a small amount of payoff to player 1 under $\mathbf{c}_1$, and player 1 transfers a small amount of payoff to player 3 under $\mathbf{c}_3$.

Theorem 2.9 implies that even if we assume that x_{ji} are unconstrained, if there is a solution to LP (2.6) with a value of $v^*(\mathbf{e}^N)$, then there is a solution with no negative payoffs from coalitions, i.e., an optimal solution that is in $I(CS)$. In particular, this means that it is no loss of generality to assume that the variables x_{ji} in LP (2.6) are unconstrained, and therefore Theorem 2.8 holds.

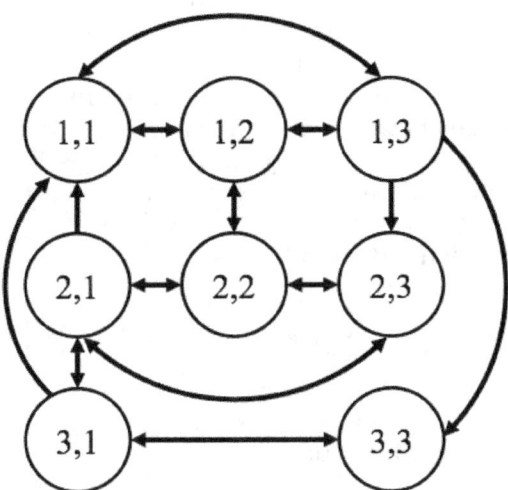

Figure 2.1: The graph $\Gamma(\mathbf{x})$ formed for the coalition structure and the imputation described in Example 2.10.

We will now provide an alternative characterization of OCF games with a non-empty conservative core, by establishing a connection between the conservative core of an OCF game and the core of a related classic cooperative game.

Given an OCF game $\mathcal{G} = \langle N, v \rangle$, its *discrete superadditive cover* is a classic cooperative game $\bar{\mathcal{G}} = \langle N, U_v \rangle$, where $U_v(S) = v^*(\mathbf{e}^S)$. Simply put, the value of a set $S \subseteq N$ in $\bar{\mathcal{G}}$ is the most that it can make under the function v by forming an overlapping coalition structure. We will now show that the conservative core of $\mathcal{G}$ is non-empty if and only if the core of $\bar{\mathcal{G}}$ is non-empty. We can, in fact, prove this in two ways: the first proof employs a graph coloring argument, as was previously used in Theorem 2.3; the second uses the characterization shown in Chalkiadakis et al. [2010], as given in Theorem 2.8.

Theorem 2.11. *The conservative core of an OCF-game $\mathcal{G} = \langle N, v \rangle$ is non-empty if and only if the core of its discrete superadditive cover $\bar{\mathcal{G}} = \langle N, U_v \rangle$ is non-empty.*

Proof via Graph Coloring. We have argued that if the conservative core of $\mathcal{G}$ is non-empty, then there exists an outcome $(CS, \mathbf{x})$ such that for all $S \subseteq N$ we have $p_S(CS, \mathbf{x}) \geq v^*(\mathbf{e}^S)$. Note that CS is an optimal coalition structure for $\mathcal{G}$, since $v(CS) = p_N(CS, \mathbf{x}) \geq v^*(\mathbf{e}^N)$. Consider the payoff vector $\mathbf{p} = (p_1, \ldots, p_n)$ where $p_i = p_i(CS, \mathbf{x})$ for all $i \in N$. It follows that $\mathbf{p}$ is in the core of $\bar{\mathcal{G}}$. Since CS is optimal, we have $\sum_{i=1}^n p_i = v(CS) = v^*(\mathbf{e}^N) = U_v(N)$, and for every $S \subseteq N$ we have $p(S) = p_S(CS, \mathbf{x}) \geq v^*(\mathbf{e}^S) = U_v(S)$.

Conversely, let $\mathbf{p} = (p_1, \ldots, p_n)$ be a payoff vector in the core of $\bar{\mathcal{G}}$. Fix a coalition structure CS such that $v(CS) = v^*(\mathbf{e}^N)$. We will now use $\mathbf{p}$ to construct an imputation $\mathbf{x} \in I(CS)$ such that $(CS, \mathbf{x})$ is in the conservative core of $\mathcal{G}$.

Given an imputation $\mathbf{z} \in I(CS)$, we color the players in N as follows: i is green if $p_i(CS, \mathbf{z}) > p_i$; red if $p_i(CS, \mathbf{z}) < p_i$ and white if $p_i(CS, \mathbf{z}) = p_i$. Green players are better off under $(CS, \mathbf{z})$ than under $\mathbf{p}$, red players are worse off under $(CS, \mathbf{z})$ than under $\mathbf{p}$, and white players are indifferent.

Now, suppose that there is some $\mathbf{z} \in I(CS)$ such that no player in the respective coloring is green. Since CS is optimal, this means that no player is red either. Thus, all players are white. This in turn implies that for all $S \subseteq N$ we have $p_S(CS, \mathbf{z}) = p(S) \geq U_v(S) = v^*(\mathbf{e}^S)$, so $(CS, \mathbf{z})$ is in the conservative core. Thus, all we need to show is that there exists some $\mathbf{z} \in I(CS)$ such that no player in the respective coloring is green.

Let $F: I(CS) \to \mathbb{R}$ be defined as follows:

$$F(\mathbf{z}) = \sum_{i=1}^{n} \max\{0, p_i - p_i(CS, \mathbf{z})\}.$$

The function F measures the players' total unhappiness with the outcome $(CS, \mathbf{z})$ as compared to $\mathbf{p}$. F is a continuous function and $I(CS)$ is a compact set. Thus, there exists an imputation $\mathbf{x} \in I(CS)$ that minimizes the value of F. We pick $\mathbf{x} \in \arg\min F(\mathbf{z})$ so as to minimize the number of white players in the respective coloring of N.

Now suppose that there exist a green player g, a non-green player i and a coalition $\mathbf{c} \in CS$ such that $g, i \in supp(\mathbf{c})$ and $x_g(\mathbf{c}) > 0$. Then there is some $\varepsilon > 0$ such that setting $\bar{x}_g(\mathbf{c}) = x_g(\mathbf{c}) - \varepsilon$ and $\bar{x}_i(\mathbf{c}) = x_i(\mathbf{c}) + \varepsilon$ (i.e., transferring an amount of ε from g to i within the coalition $\mathbf{c}$) results in a valid payoff division for CS and keeps g green. If i was white prior to this transfer, it now becomes green (and the value of F does not change). This is a contradiction to the fact that $\mathbf{x}$ minimizes the number of white players. On the other hand, if i was red, the transfer decreases the contribution of i to F (while g's contribution to F remains 0), which is a contradiction to the fact that $\mathbf{x} \in \arg\min F(\mathbf{z})$. This means that under $\mathbf{x}$, green players get zero payoff from coalitions with non-green players. Let us denote the set of green players by S_g. We have

$$p_{S_g}(CS, \mathbf{x}) = p_{S_g}(CS|_{S_g}, \mathbf{x}) \leq v^*(\mathbf{e}^{S_g}).$$

On the other hand, if $S_g \neq \emptyset$, since for each $i \in S_g$ we have $p_i(CS, \mathbf{x}) > p_i$, we obtain

$$p_{S_g}(CS, \mathbf{x}) > p(S_g) \geq U_v(S_g) = v^*(\mathbf{e}^{S_g}).$$

This contradiction shows that $S_g = \emptyset$, and we have argued that in this case all players are white. This completes the proof. □

Proof via Linear Programming Duality. First, if the c-core of $\mathcal{G}$ is not empty, then the core of the discrete superadditive cover is not empty as well. This can be shown by setting $p_i(CS, \mathbf{x}) = p_i$ for some $(CS, \mathbf{x})$ in the c-core of $\mathcal{G}$, in exactly the same manner as the graph coloring proof.

For the other direction, we show that if the discrete superadditive cover of $\mathcal{G}$ has a non-empty core, then for any optimal coalition structure CS, there exists some $\mathbf{x} \in I(CS)$ such that $(CS, \mathbf{x})$ is in the c-core of $\mathcal{G}$. Given an OCF game $\mathcal{G} = \langle N, v \rangle$ and an optimal coalition structure $CS \in \mathcal{CS}(N)$, Theorem 2.8 states that there exists an outcome $\mathbf{x} \in I(CS)$ such that $(CS, \mathbf{x})$ is in the conservative core of

$\mathcal{G}$ if and only if for every collection of non-negative weights $\{(r_j)_{j=1}^m; (\delta_S)_{S\subseteq N}\}$ that is c-balanced with respect to CS it holds that

$$\sum_{j=1}^m r_j v(\mathbf{c}_j) + \sum_{S\subseteq N} \delta_S v^*(\mathbf{e}^S) \leq v^*(\mathbf{e}^N).$$

This fact is derived using LP duality. Given an optimal coalition structure $CS = (\mathbf{c}_1, \ldots, \mathbf{c}_m)$, there is an imputation $\mathbf{x} \in I(CS)$ such that $(CS, \mathbf{x})$ is in the c-core if and only if LP (2.7) has a solution that has a value of at most $v^*(\mathbf{e}^N)$.

A point that satisfies the constraints in LP (2.7) is a c-balanced collection of weights. Given a c-balanced collection of weights $((r_j)_{j=1}^m, (\delta_S)_{S\subseteq N})$, let us write $d(i) = \sum_{S:i\in S} \delta_S$. By the c-balancedness condition, we have that for all $\mathbf{c}_j \in CS$ and for all $i \in supp(\mathbf{c}_j)$ we have $r_j + d(i) = 1$; that is, if $supp(\mathbf{c}_j) \cap supp(\mathbf{c}_{j'}) \neq \emptyset$ then $r_j = r_{j'}$.

Let us partition CS into disjoint coalition structures in the following manner. We define an undirected graph $\Gamma = \langle V, E\rangle$ where $V = \{\mathbf{c}_1, \ldots, \mathbf{c}_m\}$ and the edge $\{\mathbf{c}_j, \mathbf{c}_{j'}\}$ is in E if and only if $supp(\mathbf{c}_j) \cap supp(\mathbf{c}_{j'}) \neq \emptyset$. Let us denote the connected components of Γ by $CS_1, \ldots, CS_r$; for every two coalitions $\mathbf{c}_j, \mathbf{c}_{j'}$ in a connected component we have that $r_j = r_{j'}$. We set ρ_q to be the value of r_j for all coalitions in the connected component CS_q; since $r_j = r_{j'}$ for all $\mathbf{c}_j, \mathbf{c}_{j'} \in CS_q$, we set $\rho_q = r_j$ for all $\mathbf{c}_j \in CS_q$.

Next, for every connected component CS_q, let us write $S_q = \bigcup_{\mathbf{c}\in CS_q} supp(\mathbf{c})$. Note that if CS_p, CS_q are connected components of Γ then $S_p \cap S_q = \emptyset$; therefore, $S_1, \ldots, S_r$ are a partition of N, and $w(CS_q) = \mathbf{e}^{S_q}$ for all $q = 1, \ldots, r$. Since CS is an optimal coalition structure, it must be that for all $q = 1, \ldots, r$ we have that $v(CS_q) = v^*(\mathbf{e}^{S_q})$; therefore, under the new notation we have that

$$\sum_{j=1}^m r_j v(\mathbf{c}_j) = \sum_{q=1}^r \rho_q v(CS_q)$$
$$= \sum_{q=1}^r \rho_q v^*(\mathbf{e}^{S_q})$$

This means that given a c-balanced collection $((r_j)_{j=1}^m, (\delta_S)_{S\subseteq N})$, we can define a balanced collection of weights $(\bar\delta_S)_{S\subseteq N}$ where

$$\bar\delta_S = \begin{cases} \delta_S & \text{if } S \neq S_q \text{ for all } q \in \{1,\ldots,r\} \\ \delta_S + \rho_q & \text{if } S = S_q \text{ for some } q \in \{1,\ldots,r\} \end{cases}$$

We claim that $(\bar\delta_S)_{S\subseteq N}$ is indeed a balanced collection of weights (in the classic cooperative game sense given in [Shapley, 1967]). We need to show that for all $i \in N$, $\sum_{S:i\in S} \bar\delta_S = 1$, and that $\bar\delta_S \geq 0$. Since $S_1, \ldots, S_r$ partition N, there is a unique set S_q such that $i \in S_q$; therefore

$$\sum_{S:i \in S} \bar{\delta}_S = \bar{\delta}_{S_q} + \sum_{S:i \in S, S \neq S_q} \bar{\delta}_S$$
$$= \delta_{S_q} + \rho_q + \sum_{S:i \in S, S \neq S_q} \delta_S$$
$$= \rho_q + \sum_{S:i \in S} \delta_S = 1$$

where the last equality holds since $((r_j)_{j=1}^m, (\delta_S)_{S \subseteq N})$ is a c-balanced collection of weights, and there exists some $c_j \in CS_q$ such that $r_j = \rho_q$ and $i \in supp(c_j)$. Next, $\bar{\delta}_S \geq 0$ since both $\rho_q \geq 0$ and $\delta_S \geq 0$.

The discrete superadditive cover of $\mathcal{G}$ has a non-empty core if and only if for any collection of balanced weights $(\gamma_S)_{S \subseteq N}$, we have that

$$\sum_{S \subseteq N} \gamma_S U_v(S) \leq U_v(N).$$

Suppose that the discrete superadditive cover of $\mathcal{G}$ has a non-empty core. Let $((r_j)_{j=1}^m, (\delta_S)_{S \subseteq N})$ be any c-balanced collection of weights; we define $(\bar{\delta}_S)_{S \subseteq N}$ as above. As we have shown, $(\bar{\delta}_S)_{S \subseteq N}$ is a balanced collection of weights. We conclude that

$$\sum_{j=1}^m r_j v(\mathbf{c}_j) + \sum_{S \subseteq N} \delta_S v^*(\mathbf{e}^S) = \sum_{q=1}^r \rho_q v^*(\mathbf{e}^{S_q}) + \sum_{S \subseteq N} \delta_S v^*(\mathbf{e}^S)$$
$$= \sum_{S \subseteq N} \bar{\delta}_S v^*(\mathbf{e}^S) \leq v^*(\mathbf{e}^N),$$

where the last inequality is derived from the fact that $(\bar{\delta}_S)_{S \subseteq N}$ is a balanced collection of weights. We conclude that if the core of the discrete superadditive cover is not empty, then every c-balanced collection of weights satisfies the condition given in Theorem 2.8, and thus there exists an imputation $\mathbf{x} \in I(CS)$ such that $(CS, \mathbf{x})$ is in the conservative core of $\mathcal{G}$. □

We summarize the results we have obtained so far in the following theorem

Theorem 2.12. *Given an OCF game $\mathcal{G}$, the following are equivalent*

1. *There is an optimal coalition structure CS, such that LP (2.6) has a solution whose value is $v^*(\mathbf{e}^N)$.*

2. *The conservative core of $\mathcal{G}$ is not empty.*

3. *$\mathcal{G}$ is c-balanced with respect to some CS.*

4. *The core of the discrete superadditive cover of $\mathcal{G}$ is not empty.*

Proof. (1 ⇒ 2) is implied by Theorem 2.9. (2 ⇒ 1) holds since any imputation $\mathbf{x} \in I(CS)$ such that $(CS, \mathbf{x})$ is in the c-core satisfies the constraints of LP (2.6); LP (2.6) has a minimal value of $v^*(\mathbf{e}^N)$, which is also the value of $\mathbf{x}$ if $(CS, \mathbf{x})$ is in the c-core. (1 ⟺ 3) is shown in [Chalkiadakis et al., 2010], and (2 ⟺ 4) is shown in Theorem 2.11. We mention that the first proof of Theorem 2.11 shows (4 ⇒ 2), whereas the second shows that (4 ⇒ 3), which implies (4 ⇒ 2). □

The following corollary is immediately implied by the graph coloring proof of Theorem 2.11.

Corollary 2.13. *Consider an OCF game $\mathcal{G} = \langle N, v \rangle$. For every payoff vector $\mathbf{p} = (p_1, \ldots, p_n)$ in the core of $\bar{\mathcal{G}}$ and every optimal coalition structure $CS \in \mathcal{CS}(N)$, there exists some $\mathbf{x} \in I(CS)$ such that $p_i = p_i(CS, \mathbf{x})$ for all $i \in N$.*

As a computational aside, we note that given $(p_1, \ldots, p_n)$, in order to find an imputation $\mathbf{x} \in I(CS)$ such that $p_i(CS, \mathbf{x}) = p_i$ for all $i \in N$, one must solve a system of $|CS| + n$ linear equations. Thus, if $|CS|$ is polynomial in n, one can find such an imputation in polynomial time. We mention that it is entirely possible for $|CS|$ not to be a polynomial in n, even for natural classes of OCF games, such as network flow games. We will elaborate on this point in Chapter 3.

Corollary 2.13 enables us to generalize the following well-known result for classic cooperative games. Aumann and Drèze [1974] show that if CS and CS' are optimal (non-overlapping) coalition structures for $\mathcal{G} = \langle N, u \rangle$ and $(CS, \mathbf{p})$ is in the core of $\mathcal{G}$ then $(CS', \mathbf{p})$ is also in the core of $\mathcal{G}$. In other words, if $\mathcal{CS}_{\text{opt}}$ is the set of all optimal coalition structures for $\mathcal{G}$, and I_{stab} is the set of all stable payoff divisions for $\mathcal{G}$, then the set of core outcomes for $\mathcal{G}$ is $\mathcal{CS}_{\text{opt}} \times I_{\text{stab}}$. Using Corollary 2.13, we can extend this result to OCF games under the conservative arbitration function.

Corollary 2.14. *Consider an OCF game $\mathcal{G} = \langle N, v \rangle$. For every pair of optimal coalition structures $CS, CS' \in \mathcal{CS}(N)$ and every imputation $\mathbf{x} \in I(CS)$ such that $(CS, \mathbf{x})$ is in the conservative core, there exists an imputation $\mathbf{x}' \in I(CS')$ such that $(CS', \mathbf{x}')$ is in the conservative core of $\mathcal{G}$ and $p_i(CS, \mathbf{x}) = p_i(CS', \mathbf{x}')$ for all $i \in N$.*

Convexity in OCF Games Revisited

For classic cooperative games, *convexity*, or *supermodularity*, of the characteristic function is well-known to be a sufficient condition for non-emptiness of the core [Shapley, 1971]. In more detail, recall that a classic cooperative game $\mathcal{G} = \langle N, u \rangle$ is said to be *supermodular* if for every pair of sets S, T such that $S \subseteq T \subseteq N$ and every $R \subseteq N \setminus T$ it holds that

$$u(T \cup R) - u(T) \geq u(S \cup R) - u(S).$$

Supermodular games are often referred to as convex games in the literature; however, to avoid confusion with other notions of convexity considered in this thesis, we use the term "supermodularity". Shapley [1971] proves the following result.

Theorem 2.15 (Shapley [1971]). *If a cooperative game $\mathcal{G}$ is supermodular, then its core is non-empty.*

In order to extend Theorem 2.15 to OCF games, Chalkiadakis et al. [2010] propose the following notion of convexity for OCF games.

Definition 2.16 (OCF convexity, Chalkiadakis et al. [2010]). An OCF game is *OCF-convex* if for every pair of sets S, T such that $S \subseteq T \subseteq N$, every $R \subseteq N \setminus T$, every outcome $(CS_S, \mathbf{x}_S) \in \mathcal{F}(S)$, every outcome $(CS_T, \mathbf{x}_T) \in \mathcal{F}(T)$, and every outcome $(CS_{S \cup R}, \mathbf{x}_{S \cup R}) \in \mathcal{F}(S \cup R)$ the following condition holds: if $p_i(CS_S, \mathbf{x}_S) \leq p_i(CS_{S \cup R}, \mathbf{x}_{S \cup R})$ for all $i \in S$ then there is an outcome $(CS_{T \cup R}, \mathbf{x}_{T \cup R}) \in \mathcal{F}(T \cup R)$ such that

(1) $p_i(CS_T, \mathbf{x}_T) \leq p_i(CS_{T \cup R}, \mathbf{x}_{T \cup R})$ for all $i \in T$;

(2) $p_i(CS_{S \cup R}, \mathbf{x}_{S \cup R}) \leq p_i(CS_{T \cup R}, \mathbf{x}_{T \cup R})$ for all $i \in S \cup R$.

In words, OCF convexity states that if a set $S \subseteq T$ proposes to join a set $R \subseteq N \setminus T$, and divide profits in a way that weakly benefits the members of S, the set T can offer R an outcome that would be beneficial to all members of S, all members of R, and all members of $T \setminus S$.

Chalkiadakis et al. [2010] show that OCF convexity is a sufficient condition for non-emptiness of the conservative core.

Theorem 2.17 (Theorem 3, Chalkiadakis et al. [2010]). *If an OCF game $\mathcal{G}$ is OCF-convex, then its conservative core is non-empty.*

On the other hand, combining Theorem 2.11 with Theorem 2.15, we obtain the following sufficient condition for conservative core non-emptiness.

Proposition 2.18. *Consider an OCF game $\mathcal{G}$. If $\bar{\mathcal{G}} = \langle N, U_v \rangle$ is supermodular, then $Core(\mathcal{G}, \mathcal{A}_c)$ is non-empty.*

We will now argue that Proposition 2.18 is strictly stronger than Theorem 2.17, by showing that supermodularity of $\bar{\mathcal{G}}$ is a strictly weaker condition than OCF-convexity of $\mathcal{G}$. We first show that OCF-convexity of $\mathcal{G}$ implies supermodularity of $\bar{\mathcal{G}}$. We then present an example of a game $\mathcal{G} = \langle N, v \rangle$ such that $\bar{\mathcal{G}}$ is supermodular, but $\mathcal{G}$ is not OCF-convex (Example 2.22).

To implement the first step of this plan, we first establish the following crucial proposition, which may seem counterintuitive at first sight. Its proof is based on coloring arguments similar to the ones used in the proofs of Theorems 2.3 and 2.11.

Proposition 2.19. *For every pair of sets S, T such that $S \subseteq T \subseteq N$ and every pair of optimal coalition structures $CS_S \in \mathcal{CS}(S)$, $CS_T \in \mathcal{CS}(T)$, there exist imputations $\mathbf{x} \in I(CS_S)$ and $\mathbf{y} \in I(CS_T)$ such that $p_i(CS_S, \mathbf{x}) = p_i(CS_T, \mathbf{y})$ for all $i \in S$.*

Proof. Fix the sets S and T and optimal coalition structures $CS_S \in \mathcal{CS}(S)$, $CS_T \in \mathcal{CS}(T)$. We will first establish that we can find imputations $\bar{\mathbf{x}} \in I(CS_S)$ and $\bar{\mathbf{y}} \in I(CS_T)$ with $p_i(CS_S, \bar{\mathbf{x}}) \leq p_i(CS_T, \bar{\mathbf{y}})$ for all $i \in S$. We will then show how to use this fact to prove our original claim.

Lemma 2.20. *There exist imputations $\bar{\mathbf{x}} \in I(CS_S)$ and $\bar{\mathbf{y}} \in I(CS_T)$ such that $p_i(CS_S, \bar{\mathbf{x}}) \leq p_i(CS_T, \bar{\mathbf{y}})$ for all $i \in S$.*

Proof. Given a pair of imputations $\mathbf{x} \in I(CS_S)$, $\mathbf{y} \in I(CS_T)$, we color the players in T as follows: a player i is green if $i \in S$ and $p_i(CS_T, \mathbf{y}) > p_i(CS_S, \mathbf{x})$ or if $i \in T \setminus S$ and $p_i(CS_T, \mathbf{y}) > 0$; red if $i \in S$ and $p_i(CS_T, \mathbf{y}) < p_i(CS_S, \mathbf{x})$; and white otherwise.

We define a mapping $D : I(CS_S) \times I(CS_T) \to \mathbb{R}$ by setting

$$D(\mathbf{x}, \mathbf{y}) = \sum_{i \in S} \max\{0, p_i(CS_S, \mathbf{x}) - p_i(CS_T, \mathbf{y})\}.$$

The function D is continuous over a compact set, and thus attains its minimum value at some point $(\mathbf{x}, \mathbf{y}) \in I(CS_S) \times I(CS_T)$. Among all such points, pick one with the smallest number of white players and denote it by $(\bar{\mathbf{x}}, \bar{\mathbf{y}})$. Let $q_i = p_i(CS_T, \bar{\mathbf{y}})$ for $i \in T$, $p_i = p_i(CS_S, \bar{\mathbf{x}})$ for $i \in S$. We will now show that $q_i \geq p_i$ for all $i \in S$.

Assume for the sake of contradiction that this is not the case. Then $D(\bar{\mathbf{x}}, \bar{\mathbf{y}}) > 0$ and there is at least one red player in T. Since $v^*(\mathbf{e}^T) \geq v^*(\mathbf{e}^S)$, there is also a green player in T.

Now, consider a green player g and a non-green player i. Suppose that

(a) there exists a coalition $\mathbf{c} \in CS_T$ such that $g, i \in supp(\mathbf{c})$ and $\bar{y}_g(\mathbf{c}) > 0$, or

(b) there exists a coalition $\mathbf{c} \in CS_S$ such that $g, i \in supp(\mathbf{c})$ and $\bar{x}_i(\mathbf{c}) > 0$.

In case (a) we can modify $\bar{\mathbf{y}}$ by making g transfer a small amount of payoff to i, and in case (b) we can modify $\bar{\mathbf{x}}$ by making i transfer a small amount of payoff to g; we choose the transfer to be small enough that g remains green. If i was white, it now becomes green, and if i was red, this lowers its contribution to D. In both cases, we get a contradiction with our choice of $(\bar{\mathbf{x}}, \bar{\mathbf{y}})$. Thus, no such coalition $\mathbf{c}$ exists.

Let S_g denote the set of all green players in S, and let $S_{ng} = S \setminus S_g$. Under $\bar{\mathbf{x}}$, every $i \in S_{ng}$ gets no payoff from coalitions in CS_S whose support contains members of S_g. Similarly, under $\bar{\mathbf{y}}$ green players in T get no payoff from coalitions in CS_T that contain members of S_{ng}.

Now, suppose that we modify CS_T by having the players in S_{ng} abandon their existing coalitions and form the coalitions they form among themselves in CS_S instead. Denote the resulting coalition structure by CS'. We define an imputation $\bar{\mathbf{z}}$ for CS' as follows. $\bar{\mathbf{z}}$ coincides with $\bar{\mathbf{y}}$ for each $\mathbf{c}$ with $supp(\mathbf{c}) \cap S_{ng} = \emptyset$, and it coincides with $\bar{\mathbf{x}}$ for each coalition $\mathbf{c}'$ that is formed by the members of S_{ng}. The remaining coalitions in CS' involve both players in S_{ng} and players in $T \setminus S_{ng}$, and may have suffered a reduction in their value relative to CS_T after S_{ng} withdrew resources from them; for such coalitions $\bar{\mathbf{z}}$ distributes their value arbitrarily among the members of S_{ng}.

As argued above, in $(CS_S, \bar{\mathbf{x}})$ the members of S_{ng} only receive payoffs from coalitions that they fully control, so $p_i(CS', \bar{\mathbf{z}}) \geq p_i(CS_S, \bar{\mathbf{x}})$ for all $i \in S_{ng}$. Further, since at least one player in S is red, we have $p_{S_{ng}}(CS_S, \bar{\mathbf{x}}) > p_{S_{ng}}(CS_T, \bar{\mathbf{y}})$. On the other hand, in $(CS_T, \bar{\mathbf{y}})$ the members of $T \setminus S_{ng}$ receive no payoff from coalitions with members of S_{ng}, so letting S_{ng} receive all of the payoffs from those coalitions does not affect the payoff that the members of $T \setminus S_{ng}$ receive, relative to what they received in $(CS_T, \bar{\mathbf{y}})$. Thus, for all i in $T \setminus S_{ng}$ it holds that $p_i(CS', \bar{\mathbf{z}}) = p_i(CS_T, \bar{\mathbf{y}})$. We obtain

$$\begin{aligned}
v(CS') &= p_T(CS', \bar{\mathbf{z}}) \\
&= p_{T \setminus S_{ng}}(CS', \bar{\mathbf{z}}) + p_{S_{ng}}(CS', \bar{\mathbf{z}}) \\
&> p_{T \setminus S_{ng}}(CS_T, \bar{\mathbf{y}}) + p_{S_{ng}}(CS_T, \bar{\mathbf{y}}) \\
&= v(CS_T) = v^*(\mathbf{e}^T).
\end{aligned} \qquad (2.9)$$

This contradiction shows that the set of red players is empty and hence $p_i(CS_T, \bar{\mathbf{y}}) \geq p_i(CS_S, \bar{\mathbf{x}})$ for all $i \in S$. $\square$

We will now use Lemma 2.20 to complete the proof of Proposition 2.19. Define

$$P(CS_S, CS_T) = \{(\mathbf{x}, \mathbf{y}) \in I(CS_S) \times I(CS_T) \mid p_i(CS_T, \mathbf{y}) \geq p_i(CS_S, \mathbf{x}) \text{ for all } i \in S\}.$$

Lemma 2.20 implies that $P(CS_S, CS_T)$ is not empty. Moreover, $P(CS_S, CS_T)$ is compact; thus, it contains a point $(\mathbf{x}, \mathbf{y})$ that minimizes the value $p_S(CS_T, \mathbf{y})$ over $P(CS_S, CS_T)$.

Let $M(CS_S, CS_T)$ be the set of points $(\mathbf{x}, \mathbf{y})$ in $P(CS_S, CS_T)$ that minimize $p_S(CS_T, \mathbf{y})$; to complete the proof, we will identify a point $(\hat{\mathbf{x}}, \hat{\mathbf{y}})$ in $M(CS_S, CS_T)$ such that $p_S(CS_T, \hat{\mathbf{y}}) = p_S(CS_S, \hat{\mathbf{x}}) = v^*(\mathbf{e}^S)$.

Given a point $(\mathbf{x}, \mathbf{y}) \in M(CS_S, CS_T)$, we color the players in T as follows: i is green if $i \in S$ and $p_i(CS_T, \mathbf{y}) > p_i(CS_S, \mathbf{x})$, and white otherwise. Note that, since $(\mathbf{x}, \mathbf{y}) \in P(CS_S, CS_T)$, if S contains a white player i, we have $p_i(CS_T, \mathbf{y}) = p_i(CS_S, \mathbf{x})$. Let $(\hat{\mathbf{x}}, \hat{\mathbf{y}})$ be a point in $M(CS_S, CS_T)$ that maximizes the number of green players in T.

Now, consider a green player g and a white player i. Suppose that

(a) $g \in S$, $i \in T$, and there is a coalition $\mathbf{c} \in CS_T$ such that $g, i \in supp(\mathbf{c})$ and $y_g(\mathbf{c}) > 0$, or

(b) $g, i \in S$, and there is a coalition $\mathbf{c} \in CS_S$ such that $g, i \in supp(\mathbf{c})$ and $x_i(\mathbf{c}) > 0$.

In case (a) we could transfer a small amount of payoff from g to i in CS_T so as to either make i green and thereby increase the number of green vertices (if $i \in S$) or lower the total payoff of S in CS_T (if $i \in T \setminus S$), and in case (b) we could transfer a small amount of payoff from i to g in CS_S so as to make i green (again, we choose the transfer to be small enough that g remains green). In both cases, we get a contradiction with our choice of $(\hat{\mathbf{x}}, \hat{\mathbf{y}})$.

Let S_g be the set of all green players in S, and let $S_w = S \setminus S_g$ be the set of all white players in S. We have shown that in $(CS_S, \hat{\mathbf{x}})$ the members of S_w do not receive payoffs from their joint coalitions with members of S_g, and in $(CS_T, \hat{\mathbf{y}})$ the members of S_g do not receive payoffs from their joint coalitions with members of S_w, so we have $p_{S_w}(CS_S, \hat{\mathbf{x}}) \leq v^*(\mathbf{e}^{S_w})$ and $p_{S_g}(CS_T, \hat{\mathbf{y}}) \leq v^*(\mathbf{e}^{S_g})$. Thus, if $S_g \neq \emptyset$, we have

$$\begin{aligned}
v^*(\mathbf{e}^S) &= p_S(CS_S, \hat{\mathbf{x}}) \\
&= p_{S_g}(CS_S, \hat{\mathbf{x}}) + p_{S_w}(CS_S, \hat{\mathbf{x}}) \\
&< p_{S_g}(CS_T, \hat{\mathbf{y}}) + v^*(\mathbf{e}^{S_w}) \\
&\leq v^*(\mathbf{e}^{S_g}) + v^*(\mathbf{e}^{S_w}).
\end{aligned} \qquad (2.10)$$

Thus, if $S_g \ne \emptyset$, then $v^*(\mathbf{e}^S) < v^*(\mathbf{e}^{S_g}) + v^*(\mathbf{e}^{S_w})$, a contradiction with the superadditivity of v^*. We conclude that $S_g = \emptyset$, which implies that $p_i(CS_S, \hat{\mathbf{x}}) = p_i(CS_T, \hat{\mathbf{y}})$ for all $i \in S$. □

Armed with Proposition 2.19, we are now ready to prove the following theorem.

Theorem 2.21. *If $\mathcal{G} = \langle N, v \rangle$ is OCF-convex then $\bar{\mathcal{G}} = \langle N, U_v \rangle$ is supermodular.*

Proof. Consider sets $S, T, R \subseteq N$ such that $S \subseteq T$ and $R \subseteq N \setminus T$; we will demonstrate that $U_v(S \cup R) - U_v(S) \le U_v(T \cup R) - U_v(T)$. Set $S' = S \cup R$, $T' = T \cup R$, and consider coalition structures $CS_S \in \mathcal{CS}(S)$, $CS_{S'} \in \mathcal{CS}(S')$ such that $v(CS_S) = v^*(\mathbf{e}^S)$, $v(CS_{S'}) = v^*(\mathbf{e}^{S'})$. By Proposition 2.19, there exist imputations $\mathbf{x}_S$ and $\mathbf{x}_{S'}$ such that $p_i(CS_S, \mathbf{x}_S) = p_i(CS_{S'}, \mathbf{x}_{S'})$ for all $i \in S$; thus, the total payoff to R from $(CS_{S'}, \mathbf{x}_{S'})$ is $v^*(\mathbf{e}^{S \cup R}) - v^*(\mathbf{e}^S)$.

Consider an outcome $(CS_T, \mathbf{x}_T) \in \mathcal{F}(T)$ such that $v(CS_T) = v^*(\mathbf{e}^T)$; since $\mathcal{G}$ is OCF-convex, there is an outcome $(CS_{T'}, \mathbf{x}_{T'})$ that is better for all members of S' than $(CS_{S'}, \mathbf{x}_{S'})$, and also pays T a total of at least $v^*(\mathbf{e}^T)$. The payoff to R under $(CS_{T'}, \mathbf{x}_{T'})$ is $v(CS_{T'}) - p_T(CS_{T'}, \mathbf{x}_{T'})$. We have $v(CS_{T'}) \le v^*(\mathbf{e}^{T \cup R})$ and $p_T(CS_{T'}, \mathbf{x}_{T'}) \ge v^*(\mathbf{e}^T)$, so the payoff to R in $(CS_{T'}, \mathbf{x}_{T'})$ is at most $v^*(\mathbf{e}^{T \cup R}) - v^*(\mathbf{e}^T)$. Finally, OCF convexity of $\mathcal{G}$ implies that the payoff to R in $(CS_{T'}, \mathbf{x}_{T'})$ is at least as large as its payoff in $(CS_{S'}, \mathbf{x}_{S'})$. Combining these inequalities, we get

$$\begin{aligned} U_v(S \cup R) - U_v(S) &= v^*(\mathbf{e}^{S \cup R}) - v^*(\mathbf{e}^S) \\ &= p_R(CS_{S'}, \mathbf{x}_{S'}) \\ &\le p_R(CS_{T'}, \mathbf{x}_{T'}) \\ &\le v^*(\mathbf{e}^{T \cup R}) - v^*(\mathbf{e}^T) \\ &= U_v(T \cup R) - U_v(T), \end{aligned}$$

which concludes the proof. □

Theorem 2.21 shows that if an OCF game is OCF-convex, then its discrete superadditive cover is supermodular. We will now show that the converse does not always hold, i.e., supermodularity is a strictly weaker property than OCF convexity.

Example 2.22. Consider a game $\mathcal{G} = \langle N, v \rangle$ where $N = \{1, 2, 3\}$ and v is defined as follows:

- $v\begin{pmatrix}1\\0\\0\end{pmatrix} = 1$, $v\begin{pmatrix}0\\1\\0\end{pmatrix} = 1$;
- $v\begin{pmatrix}1\\.5\\0\end{pmatrix} = 6$, $v\begin{pmatrix}0\\.5\\1\end{pmatrix} = 3$;
- $v\begin{pmatrix}1\\0\\1\end{pmatrix} = 4$, $v\begin{pmatrix}0\\1\\1\end{pmatrix} = 4$;
- $v(\mathbf{c}) = 0$ for any other partial coalition $\mathbf{c} \in [0, 1]^3$.

We have

- $U_v(\{1\}) = U_v(\{2\}) = 1$, $U_v(\{3\}) = 0$;
- $U_v(\{1,2\}) = 6$, $U_v(\{1,3\}) = U_v(\{2,3\}) = 4$;
- $U_v(N) = 9$.

One can check that U_v is indeed supermodular. However, $\mathcal{G}$ is not OCF-convex. Set $S = \{3\}$, $T = \{1,3\}$, $R = \{2\}$, with $CS_S = \left(\binom{0}{0}{1}\right)$, $CS_T = \left(\binom{1}{0}{1}\right)$, $CS_{S\cup R} = \left(\binom{0}{1}{1}\right)$. Assume that the players in T and $S \cup R$ share the payoffs according to $\mathbf{x}_T = ((0,0,4))$ and $\mathbf{x}_{S\cup R} = ((0,4,0))$, respectively. For $\mathcal{G}$ to be OCF-convex, there has to exist a coalition structure CS where player 3 earns at least 4 and player 2 earns at least 4. However, this is impossible: if player 3 earns at least 4, then CS contains either $\mathbf{c} = \binom{1}{0}{1}$ or $\mathbf{c}' = \binom{0}{1}{1}$. If $\mathbf{c}$ is formed, then player 2 can get at most 1; if $\mathbf{c}'$ is formed, then players 2 and 3 together can get at most 4. Thus, there is no way to satisfy all of the players' demands.

2.4.2 The Sensitive Core

Recall that under the sensitive arbitration function, a player will withhold payments from the deviators if any of the coalitions it participates in are hurt by the deviation. In Section 2.3.1, we obtained the following characterization of the sensitive core: an outcome $(CS, \mathbf{x})$ is in the sensitive core if and only if CS is an optimal coalition structure, and for each set $S \subseteq N$ it holds that for each $T \subseteq N \setminus S$ the total payoff that S receives from investing its resources in coalitions involving T is at least as large as the marginal returns of investing the same resources in working on its own (see Formula (2.2)).

Let $CS = (\mathbf{c}_1, \ldots, \mathbf{c}_m)$ be an optimal coalition structure. Using the characterization above, we conclude that deciding whether the sensitive core contains an outcome of the form $(CS, \mathbf{x})$ is equivalent to determining whether the value of the following linear program equals $v^*(e^N)$.

$$\min \quad \sum_{j=1}^{m} \sum_{i \in supp(\mathbf{c}_j)} x_{ji} \qquad (2.11)$$

$$\text{s.t.} \quad \sum_{i \in supp(\mathbf{c}_j)} x_{ji} \geq v(\mathbf{c}_j) \qquad \forall \mathbf{c}_j \in CS$$

$$p_S(CS_T, \mathbf{x}) \geq v^*(\mathbf{w}_S(CS|_S \cup CS_T)) - v^*(\mathbf{w}_S(CS|_S)) \quad \forall S \subseteq N,$$
$$\forall T \subseteq N \setminus S$$

In LP (2.11) we also assume that the variables x_{ji} are unconstrained. However, since any optimal solution of LP (2.11) whose value is $v^*(e^N)$ also satisfies the constraints of LP (2.6), Theorem 2.9 ensures that if there is a solution to LP (2.11), then there is a solution where all variables have a non-negative value.

Consider the dual of this linear program.

$$\max \quad \sum_{j=1}^{m} r_j v(\mathbf{c}_j) \tag{2.12}$$
$$+ \sum_{S \subseteq N, T \subseteq N \setminus S} \delta_{S,T}(v^*(\mathbf{w}_S(CS|_S \cup CS_T)) - v^*(\mathbf{w}_S(CS|_S)))$$

s.t.
$$r_j + \sum_{S: i \in S} \sum_{\substack{T \subseteq N \setminus S: \\ supp(\mathbf{c}_j) \cap T \neq \emptyset}} \delta_{S,T} = 1 \qquad \forall \mathbf{c}_j \in CS, \forall i \in supp(\mathbf{c}_j)$$

$$r_j \geq 0 \qquad \forall \mathbf{c}_j \in CS$$
$$\delta_{S,T} \geq 0 \qquad \forall S \subseteq N, \forall T \subseteq N \setminus S$$

We say that a collection of non-negative weights $\{(r_j)_{j=1}^m; (\delta_{S,T})_{S \subseteq N; T \subseteq N \setminus S}\}$ is *s-balanced with respect to* $CS = (\mathbf{c}_1, \ldots, \mathbf{c}_m)$ if $r_j + \sum_{S:i \in S} \sum_{T \subseteq N \setminus S: supp(\mathbf{c}_j) \cap T \neq \emptyset} \delta_{S,T} = 1$ for all $\mathbf{c}_j \in CS$ and all $i \in supp(\mathbf{c}_j)$. Applying linear programming duality to linear programs (2.11) and (2.12), we obtain the following theorem.

Theorem 2.23. *The sensitive core of a game* $\mathcal{G} = \langle N, v \rangle$ *is not empty if and only if there exists a coalition structure* $CS = (\mathbf{c}_1, \ldots, \mathbf{c}_m)$ *with* $v(CS) = v^*(\mathbf{e}^N)$ *such that for every collection of non-negative weights* $\{(r_j)_{j=1}^m; (\delta_{S,T})_{S \subseteq N; T \subseteq N \setminus S}\}$ *that is s-balanced with respect to* CS *it holds that*

$$\sum_{j=1}^{m} r_j v(\mathbf{c}_j) + \sum_{S \subseteq N} \sum_{T \subseteq N \setminus S} \delta_{S,T}(v^*(\mathbf{w}_S(CS|_S) + \mathbf{w}_S(CS_T)) - v^*(\mathbf{w}_S(CS|_S))) \leq v^*(\mathbf{e}^N).$$

Theorem 2.23 is similar to Theorem 2.8, which presents a criterion for non-emptiness of the conservative core. The main difference is that Theorem 2.8 considers collections of weights that contain a single weight δ_S for each set $S \subseteq N$, whereas for the sensitive core we consider collections of weights that contain a weight $\delta_{S,T}$ for each pair of disjoint sets $S, T \subseteq N$.

We also remark that Theorem 2.23 strongly relies on the fact that the variables x_{ji} are unconstrained. That is, taken in itself, Theorem 2.23 states that there exists an s-stable *preimputation* $\mathbf{x} \in I_{pre}(CS)$ if and only if $\mathcal{G}$ is s-balanced. However, using Theorem 2.9 we can show that if there exists an s-stable preimputation $\mathbf{x} \in I_{pre}(CS)$, then there exists an s-stable imputation $\bar{\mathbf{x}} \in I(CS)$ such that $p_i(CS, \mathbf{x}) = p_i(CS, \bar{\mathbf{x}})$. This is because any s-stable preimputation must at the very least satisfy LP (2.6).

2.4.3 The Refined Core

Under the sensitive arbitration function, the deviating players only need to decide which of the non-deviating players they do not wish to work with any more. In contrast, under the refined arbitration function, the deviators have to look at the coalitions with the non-deviators one by one, and decide which of them are worth keeping. We will now provide a characterization of games with non-empty refined core that is similar to the characterizations of games with non-empty conservative and sensitive cores (Theorem 2.8 and Theorem 2.23, respectively).

Employing this characterization, we then describe a class of OCF games with a non-empty refined core.

In Section 2.3.1, we obtained the following characterization of outcomes in the refined core: given a coalition structure CS, there is an imputation $\mathbf{x} \in I(CS)$ such that $(CS, \mathbf{x})$ is in the refined core if and only if $p_S(CS', \mathbf{x}) \geq v^*(\mathbf{w}_S(CS'))$ for every $S \subseteq N$ and every coalition structure $CS' \subseteq CS$ containing $CS|_S$.

In what follows, given a subset of players $S \subseteq N$ and a coalition structure CS, we write $[CS]_S = \{CS' \subseteq CS \mid CS|_S \subseteq CS'\}$. Given a coalition structure $CS = (\mathbf{c}_1, \ldots, \mathbf{c}_m)$, a collection of non-negative weights $\{(r_j)_{j=1}^m; (\delta_{S,CS'})_{S \subseteq N; CS' \in [CS]_S}\}$ is r-balanced with respect to CS if for each $\mathbf{c}_j \in CS$ and each $i \in supp(\mathbf{c}_j)$ it holds that

$$r_j + \sum_{S: i \in S} \sum_{\substack{CS' \in [CS]_S: \\ \mathbf{c}_j \in CS'}} \delta_{S, CS'} = 1.$$

We are now ready to present our characterization.

Theorem 2.24. *The refined core of an OCF game $\mathcal{G} = \langle N, v \rangle$ is non-empty if and only if there exists a coalition structure $CS = (\mathbf{c}_1, \ldots, \mathbf{c}_m)$ with $v(CS) = v^*(\mathbf{e}^N)$ such that for every collection of weights $\{(r_j)_{j=1}^m; (\delta_{S,CS'})_{S \subseteq N; CS' \in [CS]_S}\}$ that is r-balanced with respect to CS it holds that*

$$\sum_{j=1}^m r_j v(\mathbf{c}_j) + \sum_{S \subseteq N} \sum_{CS' \in [CS]_S} \delta_{S, CS'} v^*(\mathbf{w}_S(CS')) \leq v^*(\mathbf{e}^N).$$

Proof. Fix a coalition structure $CS = (\mathbf{c}_1, \ldots, \mathbf{c}_m)$ with $v(CS) = v^*(\mathbf{e}^N)$, and consider the following linear program.

$$\begin{aligned} \min \quad & \sum_{j=1}^m \sum_{i \in supp(\mathbf{c}_j)} x_{ji} & & (2.13) \\ \text{s.t.} \quad & \sum_{i \in supp(\mathbf{c}_j)} x_{ji} \geq v(\mathbf{c}_j) & & \forall \mathbf{c}_j \in CS \\ & p_S(CS', \mathbf{x}) \geq v^*(\mathbf{w}_S(CS')) & & \forall S \subseteq N, \forall CS' \in [CS]_S \end{aligned}$$

We claim that the refined core of $\mathcal{G}$ contains an outcome of the form $(CS, \mathbf{x})$ if and only if the value of this linear program is $v^*(e^N)$. This claim would be immediate if the linear program (2.13) contained a constraint $x_{ji} \geq 0$ for each $\mathbf{c}_j \in CS$ and each $i \in supp(\mathbf{c}_j)$. We can use Theorem 2.9 to ensure that if a solution to LP (2.13) with a value of $v^*(\mathbf{e}^N)$ exists, then there is a solution with non-negative variables. However, non-negativity of x_{ji} is implied by the stability constraints. To see this, consider a coalition $\mathbf{c}_j$ such that $i \in supp(\mathbf{c}_j)$. If $supp(\mathbf{c}_j) = \{i\}$, $x_{ji} = v(\mathbf{c}_j) \geq 0$ and we are done. Otherwise, consider the coalition structure $CS' = (CS|_{\{i\}}, \mathbf{c}_j)$. The total payoff to i from CS' is $p_i(CS', \mathbf{x}) = p_i(CS|_{\{i\}}, \mathbf{x}) + x_{ji} = v(CS|_{\{i\}}) + x_{ji}$. The constraint that corresponds to $S = \{i\}$ and CS' states that this payoff must be at least $v^*(\mathbf{w}_{\{i\}}(CS'))$, which is at least $v(CS|_{\{i\}})$, and hence $x_{ji} \geq 0$.

Consider now the dual of LP (2.13).

$$\max \sum_{j=1}^{m} r_j v(\mathbf{c}_j) + \sum_{\substack{S \subseteq N \\ CS' \in [CS]_S}} \delta_{S,CS'} v^*(\mathbf{w}_S(CS')) \qquad (2.14)$$

$$\text{s.t.} \quad r_j + \sum_{\substack{S: i \in S \\ CS' \in [CS]_S : \mathbf{c}_j \in CS'}} \delta_{S,CS'} = 1 \qquad \forall \mathbf{c}_j \in CS, \forall i \in supp(\mathbf{c}_j)$$

$$\delta_{S,CS'} \geq 0 \qquad \forall S \subseteq N, \forall CS' \in [CS]_S$$

$$r_j \geq 0 \qquad \forall \mathbf{c}_j \in CS$$

Observe that the dual constraints in (2.14) are equalities since x_{ji} are unconstrained in (2.13). Note also that every feasible solution of (2.14) corresponds to a collection of non-negative weights that is r-balanced with respect to CS. Our claim now follows by the standard linear programming duality argument. □

Theorem 2.24 enables us to identify an interesting class of OCF games with a non-empty refined core. Recall that a function $f : \mathbb{R}^n \to \mathbb{R}$ is *homogeneous of degree k* (or k-homogeneous) if $\alpha f(\mathbf{x}) = f(\alpha^k \mathbf{x})$ for all $\alpha > 0$. We begin with a simple lemma, showing a relation between the homogeneity of v and v^*.

Lemma 2.25. *If v is a k-homogeneous function, then v^* is k-homogeneous.*

Proof. Since we assume that v has the ECS property, then for any $\mathbf{w} \in \mathbb{R}^n$, there exists some CS such that $\mathbf{w}(CS) = \mathbf{w}$ and $v(CS) = v^*(\mathbf{w})$. Taking some $\mathbf{w} \in \mathbb{R}_+^n$ and some $\alpha > 0$, we need to show that $\alpha^k v^*(\mathbf{w}) = v^*(\alpha \mathbf{w})$. First, there exists some coalition structure CS such that $\mathbf{w}(CS) = \mathbf{w}$; thus, we obtain that

$$\alpha^k v^*(\mathbf{w}) = \alpha^k v(CS) = \sum_{\mathbf{c} \in CS} \alpha^k v(\mathbf{c})$$
$$= \sum_{\mathbf{c} \in CS} v(\alpha \mathbf{c}) = v(\alpha \cdot CS)$$

where $\alpha \cdot CS$ denotes the coalition structure CS with the resources of every coalition in CS multiplied by a factor of α. Since $\mathbf{w}(\alpha \cdot CS) = \alpha \mathbf{w}$, we know that $\alpha^k v^*(\mathbf{w}) \leq v^*(\alpha \mathbf{w})$. The other direction can be obtained in a similar fashion. Take CS' such that $v^*(\alpha \mathbf{w}) = v(CS')$; then

$$v^*(\alpha \mathbf{w}) = \sum_{\mathbf{c} \in CS'} v(\mathbf{c}) = \sum_{\mathbf{c} \in CS'} v(\alpha \frac{1}{\alpha} \mathbf{c})$$
$$= \alpha^k \sum_{\mathbf{c} \in CS'} v(\frac{1}{\alpha} \mathbf{c}) = \alpha^k v^*(\frac{1}{\alpha} CS').$$

Since $\mathbf{w}(\frac{1}{\alpha} CS') = \mathbf{w}$, we have that $v(\frac{1}{\alpha} CS') \leq v^*(\mathbf{w})$, implying that $v^*(\alpha \mathbf{w}) \leq \alpha^k v(\mathbf{w})$, which concludes the proof. □

Corollary 2.26. *Consider an OCF game $\mathcal{G} = \langle N, v \rangle$. if v or v^* is homogeneous of degree $k \geq 1$, then the refined core of $\mathcal{G}$ is non-empty.*

Proof. According to Lemma 2.25, it suffices to show the claim for v^*. Consider a coalition structure $CS = (\mathbf{c}_1, \ldots, \mathbf{c}_m)$ with $v(CS) = v^*(\mathbf{e}^N)$ and a collection of non-negative weights $\{(r_j)_{j=1}^m; (\delta_{S,CS'})_{S \subseteq N; CS' \in [CS]_S}\}$ that is r-balanced with respect to CS. According to Theorem 2.24, it suffices to show that

$$\sum_{j=1}^m r_j v(\mathbf{c}_j) + \sum_{\substack{S \subseteq N \\ CS' \in [CS]_S}} \delta_{S,CS'} v^*(\mathbf{w}_S(CS')) \leq v^*(\mathbf{e}^N). \tag{2.15}$$

We have

$$\sum_{\substack{S \subseteq N \\ CS' \in [CS]_S}} \delta_{S,CS'} v^*(\mathbf{w}_S(CS')) = \sum_{\substack{S \subseteq N \\ CS' \in [CS]_S}} v^*(\delta_{S,CS'}^k \mathbf{w}_S(CS')) \tag{2.16}$$

$$\leq \sum_{\substack{S \subseteq N \\ CS' \in [CS]_S}} v^*(\delta_{S,CS'} \mathbf{w}_S(CS'))$$

$$\leq v^*\left(\sum_{\substack{S \subseteq N \\ CS' \in [CS]_S}} \delta_{S,CS'} \mathbf{w}_S(CS')\right).$$

The first equality is due to the k homogeneity of v^*; the next inequality uses the fact that $\delta_{S,CS'} \leq 1$ and that $k \geq 1$, along with the fact that v^* is monotone; the last inequality uses the fact that v^* is superadditive.

Denote the i-th coordinate of $\sum_{\substack{S \subseteq N \\ CS' \in [CS]_S}} \delta_{S,CS'} \mathbf{w}_S(CS')$ by ω_i. We have

$$\omega_i = \sum_{S: i \in S} \sum_{CS' \in [CS]_S} \delta_{S,CS'} \sum_{\mathbf{c}_j \in CS'} c_{ji}$$

$$= \sum_{j=1}^m c_{ji} \sum_{S: i \in S} \sum_{\substack{CS' \in [CS]_S: \\ \mathbf{c}_j \in CS'}} \delta_{S,CS'}$$

$$= \sum_{j=1}^m c_{ji}(1 - r_j),$$

where the first transition is derived from the fact that $w_i(CS') = \sum_{\mathbf{c}_j \in CS'} c_{ji}$ if $i \in S$, and the second transition uses the fact that $\{(r_j)_{j=1}^m; (\delta_{S,CS'})_{S \subseteq N; CS' \in [CS]_S}\}$ is r-balanced with respect to CS. We conclude that the right-hand side of equation (2.16) is upper-bounded by

$$v^*\left(\sum_{j=1}^m (1 - r_j) \mathbf{c}_j\right).$$

We now consider the first summand in the left-hand side of (2.15). We have

$$\sum_{j=1}^{m} r_j v(\mathbf{c}_j) = \sum_{j=1}^{m} v(r_j^k \mathbf{c}_j) \quad (2.17)$$

$$\leq \sum_{j=1}^{m} v^*(r_j^k \mathbf{c}_j)$$

$$\leq \sum_{j=1}^{m} v^*(r_j \mathbf{c}_j) \leq v^*(\sum_{j=1}^{m} r_j \mathbf{c}_j)$$

The first equality holds by k-homogeneity of v; the next inequality holds since $v \leq v^*$; the next inequality holds since $r_j^k \leq r_j$ and due to the monotonicity of v^*, and the final inequality holds since v^* is superadditive. Combining (2.16) and (2.17), we obtain that the left-hand side of (2.15) can be upper-bounded by

$$v^*(\sum_{j=1}^{m} r_j \mathbf{c}_j + \sum_{j=1}^{m}(1 - r_j)\mathbf{c}_j) = v^*(\sum_{j=1}^{m} \mathbf{c}_j) = v^*(\mathbf{e}^N),$$

which concludes the proof. □

In fact, the proof of Corollary 2.26 shows a stronger claim: if v is homogeneous of degree $k \geq 1$, then *every* optimal coalition structure CS admits an imputation $\mathbf{x} \in I(CS)$ such that $(CS, \mathbf{x})$ is in the refined core. This is not the case in general; there exist games where some optimal coalition structures cannot be stabilized under the refined arbitration function, while others can. This is illustrated by the following example.

Example 2.27. Consider the following three-player game. There are four types of tasks: A task of type t_1 can be completed by player 1 alone, requires all of its resources, and is worth 5. A task of type t_{12} requires 50% of both player 2 and player 1's resources and is worth 10. A task of type T_{12} requires all of the resources of players 1 and 2 and is worth 20. Finally, a task of type t_{23} requires all of player 3's resources and 50% of player 2's resources, and is worth 9. Consider coalition structures $CS = (\mathbf{c}_1, \mathbf{c}_2)$ and $CS' = (\mathbf{c}'_1)$, where

$$\mathbf{c}_1 = \begin{pmatrix} .5 \\ .5 \\ 0 \end{pmatrix}, \mathbf{c}_2 = \begin{pmatrix} .5 \\ .5 \\ 0 \end{pmatrix}, \mathbf{c}'_1 = \begin{pmatrix} 1 \\ 1 \\ 0 \end{pmatrix}.$$

It is easy to see that both CS and CS' are optimal for this game. Simply put, it is best for players 1 and 2 to work together and earn a total of 20 (by completing t_{12} twice or T_{12} once), while player 3 makes no profit.

First, we claim that CS cannot be stabilized with respect to the refined arbitration function. The reason is that for an outcome $(CS, \mathbf{x})$ to be in the refined core, it must be the case that player 2 gets at least 9 from each of the coalitions $\mathbf{c}_1$ and $\mathbf{c}_2$. However, this means that player 1 gets at most 2 from working with player 2, while he can get 5 by working alone. On the other hand, let y be the imputation for CS' that splits the payoff from T_{12} evenly between players 1 and 2. Then $(CS', \mathbf{y})$ is in the refined core.

Example 2.27 shows that arbitration functions are not a trivial extension of the classic model of cooperative games. The results for the conservative core we have previously shown are, from a modeling point of view, rather disappointing. They essentially show that if one assumes a very simple reaction to deviation, then the arbitration model can be reduced to the classical model. However, if one assumes a more nuanced reaction to deviation, one is able to observe phenomena that do not appear in the classic model, such as the one described in Example 2.27. In more detail: under the classical (and conservative OCF) model, if one optimal coalition structure can be stabilized, then all optimal coalition structures can be stabilized as well, via the same payoff division. That is, it is essentially unimportant which coalition structure is formed, as long as it is optimal. Example 2.27 shows that the coalition structure that players choose has a dramatic effect on stability.

Example 2.27 highlights another simple structural observation: under the refined arbitration function, greater coalitional interdependency leads to greater stability. We formally capture this in the following proposition.

Proposition 2.28. *Given an outcome* $(CS, \mathbf{x}) \in \mathcal{F}(N)$, *suppose that there exist* $\mathbf{c}, \mathbf{c}' \in CS$ *such that* $v(\mathbf{c}) + v(\mathbf{c}') = v(\mathbf{c} + \mathbf{c}')$. *Let CS' be the coalition structure CS but with $\mathbf{c}, \mathbf{c}'$ replaced with* $\hat{\mathbf{c}} = \mathbf{c} + \mathbf{c}'$, *and let* $\mathbf{y} \in I(CS')$ *be an imputation that pays all players according to* $\mathbf{x}$, *and with* $\mathbf{y}(\hat{\mathbf{c}}) = \mathbf{x}(\mathbf{c}) + \mathbf{x}(\mathbf{c}')$. *If $(CS, \mathbf{x})$ is in the r-core of $\mathcal{G}$, then so is $(CS', \mathbf{y})$.*

In other words, when seeking r-stability in OCF games, it is generally advisable to choose coalition structures that focus on a small number of large projects, rather than many small, independent projects.

Corollary 2.26 shows the existence of an r-stable outcome $(CS, \mathbf{x})$ whenever CS is an optimal coalition structure, it does not immediately translate to a polynomial time algorithm for finding such an imputation. As we show in Section 3.6, such algorithms do exist for a class of 1-homogeneous games, linear bottleneck games. As we show in Chapter 5, homogeneous functions play an important role when defining other models of incentives in collaborative scenarios.

Finally, we note that one cannot immediately employ an LP duality argument for characterizing OCF games with non-empty optimistic core. The difficulty is that under the optimistic arbitration function the number of possible deviations is infinite: a deviating set needs to specify the *amount* of resources is withdraws from each coalition with non-deviators. In other words, the linear program that describes the optimistic core has infinitely many constraints, so its dual has infinitely many variables.

Chapter 3

Computing Stable Outcomes in OCF Games

Chapter 2 introduces the concept of arbitration functions, and presents a family of solution concepts for OCF games: the arbitrated core of an OCF game. In this chapter, we discuss computational aspects of OCF games.

The theory of OCF games is a generalization of classic cooperative game theory, where computational issues are relatively well-studied (see Chalkiadakis et al. [2011] for an overview). However, as Chalkiadakis et al. [2010] show, more elaborate reactions to deviation may make even tractable instances of computing solution concepts for classic cooperative games NP-hard. For example, Chalkiadakis et al. [2010] show that finding a payoff division in the c-core of threshold task games can be done in pseudopolynomial time (i.e. in time polynomial in the number of players and in the number of bits used to encode the largest weight of any player), however, under the refined arbitration function, the same problem becomes NP-hard.

To conclude, when assessing the computational complexity of finding stable outcomes in OCF games, one needs to consider structural properties of both the characteristic function and the arbitration function. In this chapter, we study some of the computational issues that arise from these two aspects of OCF games, when deciding stability related questions in OCF games.

We study several closely related questions. Given an OCF game $\mathcal{G} = \langle N, v \rangle$,

OPTVAL: compute $v^*(\mathbf{c})$ for a given $\mathbf{c} \in [0,1]^n$.

ARBVAL: compute $\mathcal{A}^*(CS, \mathbf{x}, S)$ for a given $(CS, \mathbf{x}) \in \mathcal{F}(N)$ and $S \subseteq N$.

CHECKCORE: decide whether $(CS, \mathbf{x}) \in Core(\mathcal{G}, \mathcal{A})$ for a given outcome $(CS, \mathbf{x})$ and a given arbitration function $\mathcal{A}$.

IS-STABLE: decide whether $Core(\mathcal{G}, \mathcal{A}) \neq \emptyset$.

We observe that all four problems are closely related. For example, an algorithm that solves ARBVAL can be used to solve OPTVAL, by setting $\mathcal{A} = \mathcal{A}_c$ as its input.

As we later show, an algorithm that solves CHECKCORE can be used in order to decide IS-STABLE under certain conditions.

We make two attempts to answer these four questions. In the first part of this chapter (sections 3.2 to 3.5), we make no assumptions on how players generate revenue (i.e., the characteristic function can take any form), and focus on the structure of their interaction. In Section 3.2, we show that in order to circumvent computational intractability, several conditions must be met. First, given the results of Chalkiadakis et al. [2010], we search for pseudopolynomial time algorithms that decide the above questions (foregoing this assumption results in NP-hardness, even for a single player). Second, players cannot form large coalitions; third, as shown in Section 3.3, player reaction to deviators must also be limited in its scope, as complex reactions to deviation are a source of computational complexity in themselves. Inspired by the work of Demange [2004], we study settings where player interactions have a tree structure. Under these assumptions, optimal coalition structures and profitable deviations can be found (as shown in Sections 3.2 and 3.3); moreover, deciding whether a payoff division is stable and whether the core is not empty can be done in polynomial time (see Section 3.4).

In order to provide a clearer picture of the role of interaction structure in limiting computational complexity, we provide a connection between the treewidth of the player interaction graph and the computational complexity of finding stable outcomes in OCF games. In Section 3.5, we show how the algorithms described in previous sections can be extended to algorithms whose running time is polynomial in n and $(W_M + 1)^k$, where n is the number of players, W_M is the maximal weight of any player, and k is the treewidth of the interaction graph.

In the second part of this chapter (Section 3.6), we make no assumptions on the structure of player interaction, but rather limit our attention to a specific class of OCF games. That is, instead of limiting how players interact, we limit the way in which they generate profits. More specifically, we study a class of games we term *linear bottleneck games* (LBGs); our main insight is that these games have a non-empty optimistic core; thus, if one assumes that accountability holds, LBGs have a non-empty $\mathcal{A}$-core for any $\mathcal{A}$, and an outcome in $Core(\mathcal{G}, \mathcal{A})$ can be found in polynomial time. Our results extend those of Markakis and Saberi [2005], who show that a subclass of these games (namely, multicommodity flow games) is stable in the classic cooperative sense.

3.1 Discrete OCF Games

Before we proceed with our formal analysis, let us introduce a variant of the OCF model described in Section 1.3, which we term *discrete OCF games*. A discrete OCF game is a tuple $\mathcal{G} = \langle N, \mathbf{W}, v \rangle$, where $\mathbf{W} = (W_1, \ldots, W_n)$ is a vector of positive integers; W_i is referred to as the *weight* of player i. Let us write $\mathcal{W} = \{\mathbf{c} \in \mathbb{Z}_+^n \mid \mathbf{c} \leq \mathbf{W}\}$ to be the set of all possible coalitions that players may form; then $v : \mathcal{W} \to \mathbb{R}$ is the characteristic function of the discrete OCF game $\mathcal{G}$, i.e. $v(\mathbf{c})$ is the revenue generated if for all $i \in N$, player i contributes c_i. Simply put, rather than having players contribute any fraction of their weight to a coalition, in discrete OCF games we assume that players are only allowed to contribute integer parts of their weight. The rest of the definitions concerning OCF games (coalition structures, imputations and arbitration functions) are extended naturally to discrete OCF games; this is because discrete OCF games are essentially equivalent to the OCF games considered so far, but where the characteristic function v only has positive values at finitely many rational points in $[0,1]^n$.

3.2 Computing Optimal Coalition Structures in OCF Games

We now begin our formal computational analysis of OCF games, starting with the fundamental problem of finding optimal coalition structures.

We begin by describing the formal computational model that we study in this chapter. Given a discrete OCF game $\mathcal{G} = \langle N, \mathbf{W}, v \rangle$, we assume that the value $v(\mathbf{c})$ for all $\mathbf{c} \in \mathcal{W}$ is computable in polynomial time. Moreover, we assume that the arbitration function can be computed in polynomial time for all inputs. In other words, we assume that we have oracle access to both v and $\mathcal{A}$. This assumption does not trivially hold —a naive representation of the function v is a list of $|\mathcal{W}| = \prod_{i=1}^{n}(W_i + 1)$ values, one for each possible coalition that the players may form; since $W_i \geq 1$ for all $i \in N$, it follows that $|\mathcal{W}| \geq 2^n$, i.e., it is exponential in the natural problem parameters. Succinct representation of coalitional games, be it OCF or non-OCF games, is an interesting problem in its own right; however, apart from *linear bottleneck games*, which will be the subject of discussion later in this chapter, we do not discuss representation issues in OCF games. Rather than focusing on a particular succinct representation, we make the following simplifying assumption: *the players cannot form arbitrarily large coalitions* —i.e., coalition size is bounded by a constant. Formally, we capture this notion in the following definition.

Definition 3.1 (k-OCF Games). An OCF game $\mathcal{G} = \langle N, \mathbf{W}, v \rangle$ is a k-OCF game if, for all $\mathbf{c} \in \mathcal{W}$, if $|supp(\mathbf{c})| > k$ then $v(\mathbf{c}) = 0$.

Definition 3.1 applies to several real-life scenarios where overlapping coalitions form. In many market scenarios, transactions are performed involving only few parties; in social network applications, players form pairwise coalitions; in many large-scale collaborative projects, small teams are formed to tackle various aspects, as large teams of collaborators tend to be less efficient. In order to simplify notation, given a set of players $i_1, \ldots, i_k$, let us write $v_{i_1,\ldots,i_k}(w_{i_1}, \ldots, w_{i_k})$ to denote the value of v when player i_1 contributes w_{i_1}, player i_2 contributes w_{i_2} and so on. For example, $v_{i,j}(w_i, w_j)$ is the value of players i and j collaborating, where player i contributes w_i and player j contributes w_j. We note that the number of coalitions that can have a positive value in k-OCF games is $\binom{n}{k}(W_M + 1)^k$, so if k is a constant, then the characteristic function for a k-OCF game can be represented by a polynomial number of bits.

3.2.1 Finding an optimal coalition structure

The problem of "finding an optimal coalition structure" in an OCF setting can be rephrased as follows. Given a k-OCF game $\mathcal{G} = \langle N, \mathbf{W}, v \rangle$, we are interested in the following question: is the maximal ("optimal") profit that players expect to (collectively) accumulate by forming an overlapping coalition structure greater than a given value?

Definition 3.2 (OPTVAL). Given a discrete OCF game $\mathcal{G} = \langle N, \mathbf{W}, v \rangle$, a coalition $\mathbf{c} \in \mathcal{W}$ and a value $V \in \mathbb{Q}$, an instance of OPTVAL$(\mathcal{G}, \mathbf{c}, V)$ is a "yes" instance if and only if $v^*(\mathbf{c}) \geq V$.

Letting $W_M = \max_{i \in N} W_i$, the following simple proposition is an important first step in computing solutions for OCF games.

Proposition 3.3. *If $|supp(\mathbf{c})| = m$, then* OPTVAL$(\mathcal{G}, \mathbf{c}, V)$ *is computable in time polynomial in $(W_M + 1)^m$.*

Proof. We claim that Algorithm 8 can compute an optimal coalition structure in polynomial time. To see why this is the case, we observe that

$$v^*(\mathbf{c}) = \max \{v(\mathbf{c}), \{v^*(\mathbf{c} - \mathbf{d}) + v(\mathbf{d}) \mid \mathbf{d} \leq \mathbf{c}; \mathbf{d} \neq \mathbf{c}\}\}.$$

Observe that the number of coalitions $\mathbf{d}$ such that $\mathbf{d} \leq \mathbf{c}$ is at most $(W_M + 1)^m$, so computing $v^*(\mathbf{c})$ requires computing at most $(W_M + 1)^m$ values of v^*, which require a comparison of most $(W_M + 1)^m$ values each, for a total running time that is polynomial in $(W_M + 1)^m$. □

Algorithm 1: Computing an optimal coalition structure in 2-OCF games for coalitions with a support of size m

---
1 **Algorithm** OptVal$(\mathcal{G} = \langle N, \mathbf{W}, v \rangle, \mathbf{c})$
2   **if** $\mathbf{c} = 0^n$ **then**
3   **return** 0;
4     $M \leftarrow v(\mathbf{c})$;
5   **for** $\mathbf{d} \leq \mathbf{c}; \mathbf{d} \neq \mathbf{c}$ **do**
6        $M' \leftarrow$ OptVal$(\mathcal{G}, \mathbf{d}) + v(\mathbf{c} - \mathbf{d})$;
7        $M \leftarrow \max\{M, M'\}$;
8   **return** M;

Proposition 3.3 implies that if $|supp(\mathbf{c})|$ is a constant, and W_M is polynomial in n, then OPTVAL$(\mathcal{G}, \mathbf{c}, V)$ can be computed in time polynomial in n. This result is the best one can hope for: Chalkiadakis et al. [2010] show via a reduction from the KNAPSACK problem [Garey and Johnson, 1979] that when player weights are large, OPTVAL is NP-complete. We now turn to study OPTVAL when the group of players involved is large—i.e., $|supp(\mathbf{c})|$ is not a constant. We stress that when $|supp(\mathbf{c})|$ is not a constant, this simply means that a large group of players wants to form a coalition structure. Proposition 3.3 was concerned with a small number of players wanting to find the best way to share resources among themselves; when the number of players is not small, the actual coalitions that they form may still be. For example, if $|supp(\mathbf{c})| = 500$, this means that there are 500 players who wish to find the best way for them to form smaller coalitions using the resources available as per $\mathbf{c}$, e.g. by forming pairwise interactions (if $\mathcal{G}$ is a 2-OCF game). We begin by showing the following negative result.

Proposition 3.4. OPTVAL$(\mathcal{G}, \mathbf{c}, V)$ is NP-complete even if one assumes that $\mathcal{G}$ is a 2-OCF game, and that $W_M = 3$.

Proof. First, we observe that OPTVAL is in NP; it suffices to guess a coalition structure CS such that $\mathbf{w}(CS) = \mathbf{c}$ and checking whether $v(CS) \geq V$; note that the size of this coalition structure is at most nW_M, which is polynomial in the input size, assuming that W_M is polynomial in n.

For the hardness proof, we provide a reduction from EXACT COVER BY 3-SETS (X3C) [Garey and Johnson, 1979]. Recall that an instance $\mathcal{X} = \langle A, \mathcal{S} \rangle$ of X3C is given by a finite set A, $|A| = 3\ell$, and a collection of subsets $\mathcal{S} = \{S_1, \ldots, S_t\}$ such that $S_j \subseteq A$ and $|S_j| = 3$ for all $j = 1, \ldots, t$. It is a "yes"-instance if A can be exactly covered by sets from $\mathcal{S}$; that is, if there exists a subset $\mathcal{S}' \subseteq \mathcal{S}$ such that $\bigcup_{S \in \mathcal{S}'} S = A$, and for any two $S, T \in \mathcal{S}'$ we have $S \cap T = \emptyset$.

Given an instance $\mathcal{X} = \langle A, \mathcal{S} \rangle$ of X3C, we construct a discrete 2-OCF game $\mathcal{G}(\mathcal{X}) = \langle N, v \rangle$ with $W_M = 3$ as follows. We have a player a_i of weight 1 for every element $i \in A$ and a player a_S with weight 3 for every $S \in \mathcal{S}$. The characteristic function is defined as follows: if $i \in S$, then the value of both a_i and a_S forming a coalition where each of them contributes a weight of 1 is 2; that is, if $supp(\mathbf{c}) = \{a_i, a_S\}$ and player contributions are $c_{a_i} = c_{a_S} = 1$, then $v(\mathbf{c}) = 2$. Moreover, players corresponding to sets in $\mathcal{S}$ can generate profits by working alone: if they dedicate all their weight to forming a singleton coalition, they generate a profit of 5. In other words, if $supp(\mathbf{c}) = \{a_S\}$ and $c_{a_S} = 3$ then $v(\mathbf{c}) = 5$. The value of every other partial coalition in the game $\mathcal{G}(\mathcal{X})$ is 0.

Let $S \in \mathcal{S}$ be $\{x, y, z\}$, and consider the set of players $G_S = \{a_S, a_x, a_y, a_z\}$. Collectively, the players in G_S can earn 6 if a_S forms a partial coalition with each of a_x, a_y, and a_z, and contributes one unit of weight to each of these coalitions; in any other coalition structure, G_S earns at most 5. Hence, $\mathcal{X} = \langle A, \mathcal{S} \rangle$ admits an exact cover if and only if $v^*(\mathbf{W}) \geq 6\frac{|A|}{3} + 5(t - \frac{|A|}{3}) = 5t + \frac{|A|}{3}$. On the one hand, if $\mathcal{X}$ is a "yes" instance of X3C, then there exists some $\mathcal{S}' \subseteq \mathcal{S}$ of size $\frac{|A|}{3}$ that exactly covers A. In that case,

$$v^*(\mathbf{W}) \geq \sum_{S \in \mathcal{S}'} v^*(\mathbf{W}^{G_S}) + \sum_{S \notin \mathcal{S}'} v^*(\mathbf{W}^{G_S}) = 6\frac{|A|}{3} + 5(t - \frac{|A|}{3}).$$

On the other hand, if $v^*(\mathbf{W}) \geq 6\frac{|A|}{3} + 5(t - \frac{|A|}{3})$ then this means that there is a subset of $\mathcal{S}$, $\mathcal{S}'$, of size at least $\frac{|A|}{3}$ such that for all $S \in \mathcal{S}'$ we have that $v^*(\mathbf{W}^{G_S}) = 6$. Note that since $|S| = 3$ for all $S \in \mathcal{S}'$ and since $S \cap T = \emptyset$ for all $S \in \mathcal{S}'$, it must be that $|\mathcal{S}'| = \frac{|A|}{3}$, i.e. $\mathcal{S}'$ is a partition of A; $\mathcal{S}'$ is an exact cover of A. □

Proposition 3.4 severely limits our prospects of finding tractable OPTVAL$(\mathcal{G}, \mathbf{c}, V)$ cases: even if player interactions are limited to pairwise interactions, and player weights are small constants, OPTVAL remains hard. Observe that the hardness of OPTVAL implies the hardness of most other problems of interest in our work: the most that a set can gain by deviating, the stability of a given game, and the possibility of deviation from a given outcome of a game. Thus, in order to proceed, we must first identify some limiting conditions that make OPTVAL computationally tractable.

3.2.2 Limiting Interactions in OCF Games

Demange [2004] shows that if one assumes a hierarchical player interaction structure in a cooperative game, then the core of the game is not empty; moreover, it is possible to find a core imputation in polynomial time. We now show how to adapt this interesting result to the OCF setting. The formal model we propose is not unlike the one presented by Demange. A player *interaction graph* is a graph $\Gamma = \langle N, E \rangle$, where the edges in E represent valid player interactions; given an OCF game $\mathcal{G} = \langle N, \mathbf{W}, v \rangle$, the game $\mathcal{G}$ *reduced to* Γ, denoted $\mathcal{G}|_\Gamma = \langle N, v|_\Gamma \rangle$ is defined as follows: for every $\mathbf{c} \in \mathcal{W}$, if the nodes in $supp(\mathbf{c})$ induce a connected subgraph in Γ then $v|_\Gamma(\mathbf{c}) = v(\mathbf{c})$, otherwise $v|_\Gamma(\mathbf{c}) = 0$. Such graph restrictions on player interaction are known in classic cooperative game theory as *Myerson graphs* [Myerson, 1977]. Unfortunately, limiting player interaction is not a sufficient condition for the tractability of OPTVAL. As shown in [Chalkiadakis et al., 2010], finding an optimal coalition structure is hard, even for a single player, if his weight is sufficiently large. Moreover, even if weights are small but one does not limit the size of allowable coalitions, assuming that players interact based on a hierarchical tree structure does not aid computational complexity, as shown by the following proposition.

Proposition 3.5. *OPTVAL is NP-hard, even if one is limited to the set of instances $(\mathcal{G}|_T, \mathbf{c}, V)$ where T is a tree and player weights are constants.*

Proof. Our reduction is from the INDEPENDENT-SET problem [Garey and Johnson, 1979]. An instance of INDEPENDENT-SET is a tuple $\langle \Gamma, m \rangle$, where $\Gamma = \langle N, E \rangle$ is a graph and m is an integer. It is a "yes" instance if Γ contains an independent set of size at least m, and is a "no" instance otherwise. Recall that an independent set in Γ is a subset of vertices $S \subseteq N$ such that if $i, j \in S$, then the edge $\{i, j\}$ is not in E. In other words, it is a subset of vertices that do not share an edge with one another. Given an instance $\langle \Gamma = \langle N, E \rangle, m \rangle$ of INDEPENDENT-SET, we construct the following instance of OPTVAL: we set the player set to be $N' = N \cup \{n+1\}$, and have the interaction graph T connect all vertices in N with $n+1$ via an edge. We set all players in N to have a weight of 1, whereas player $n+1$ has a weight of 2. The characteristic function v is defined as follows: given a set $S \subseteq N$, if it is an independent set in Γ, and player $n+1$ allocates a weight of 1 to working with S, then the value of the resulting coalition is 1. If S forms a vertex cover of Γ (i.e., all vertices in Γ share an edge with the vertices in S), then the value of the resulting coalition is $\frac{\varepsilon}{|S|+1}$, where ε is some constant much smaller than 1. In order to form an optimal coalition structure, $n+1$ must allocate a weight of 1 to the maximal independent set, and a weight of 1 to working with the rest of the vertices. This is because if S is an independent set, then $N \setminus S$ is a vertex cover. We conclude that the value of the optimal coalition structure in the resulting game is more than $1 + \frac{\varepsilon}{n-m+1}$ if and only if there exists an independent set of size at least m. □

Proposition 3.5 does not hold for classic cooperative games: this would be a contradiction to Demange's theorem. However, it is worth noting that the game described in Proposition 3.5 is "almost" a classic cooperative game: there is only one player whose weight is not 1, and his weight is only 2. Indeed, changing player $n+1$'s weight from 2 to 1 in the above theorem would result in a game with a trivially optimal coalition structure: forming a coalition structure $\{N' \setminus \{i, n+1\}, \{i, n+1\}\}$ for any $i \in N$ is trivially optimal, as a single vertex is always an independent set.

Propositions 3.4 and 3.5 suggest the following three conditions are *necessary* in order to find an optimal coalition structure in polynomial time:

1. Limiting player weights.

2. Limiting player interactions.

3. Limiting coalition size.

Dropping any of these three limitations results in hard instances of OPTVAL. We can, however, compute an optimal structure for games whose interaction graphs are trees in time polynomial in W_M, assuming that $\mathcal{G}$ is a 2-OCF game. Given a tree $T = \langle N, E \rangle$ whose root is some $r \in N$, and a player $i \in N$, let T_i be the subtree rooted in i, and $C_i(T)$ be the children of i in T. Given a graph Γ we refer to the nodes of Γ as $N(\Gamma)$. Finally, given a coalition $\mathbf{c} \in \mathcal{W}$, we write $(\mathbf{c}_{-i}, w)$ to be the coalition $\mathbf{c}$ with the i-th coordinate replaced with w. For ease of exposition, we say that $\mathcal{G}$ has a *tree interaction structure* if there exists some tree $T = \langle N, E \rangle$ such that $\mathcal{G} = \mathcal{G}|_T$; this way, T is part of the game description, and does not have to be an additional input.

Theorem 3.6. *Given a discrete 2-OCF game $\mathcal{G} = \langle N, \mathbf{W}, v \rangle$ such that $\mathcal{G}$ has a tree interaction structure, OPTVAL can be decided in time polynomial in $(W_M + 1)$ and n for $\langle \mathcal{G}, \mathbf{c}, V \rangle$.*

Proof. We will show that Algorithm 2 computes $v^*(\mathbf{W})$ within the required bounds; however, our result easily holds for general $\mathbf{c} \in \mathcal{W}$: simply look at the game where player weights are as per $\mathbf{c}$.

We arbitrarily choose some player $r \in N$ to be the root of the interaction graph, and process the players in N from the leaves up to the root. The key observation to make is that in order to find an optimal allocation of player resources, each node in the tree needs to decide how much weight is to be allocated to its subtree, how much is to be allocated to collaborating with its parent, and how much is to be allocated to working alone. Note that according to Proposition 3.3, both $v_i^*(x)$ and $v_{i,j}^*(x, y)$ can be computed in time polynomial in W_M. Given a node $i \in N$, let C_i be the children of i in the interaction graph. Let $v_{T_i}^*(w)$ be the most that the subtree T_i can make if player i allocates w to working with T_i. We note that

$$v_{T_i}^*(w) = \max_{\substack{y_i + \sum_{c \in C_i} w_c = w \\ \forall c \in C_i : 0 \leq x_c \leq W_c}} \{v_i^*(y_i) + \sum_{c \in C_i} v_{i,c}^*(w_c, x_c) + v_{T_c}^*(W_c - x_c)\}.$$

Using this recurrence relation we obtain the following dynamic programming method of finding an optimal coalition structure, as described in Algorithm 2 and its subroutine.

Suppose that we have already computed $v^*_{T_c}(w)$ for all $c \in C_i$ and all $w = 0, \ldots, w_c$. Now, let us write $C_i = \{c_1, \ldots, c_m\}$, and let $T_{i,j}$ be the subtree T_i, but with the subtrees $T_{c_{j+1}}, \ldots, T_{c_m}$ removed. Let $v^*_{T_{i,j}}(w)$ be the maximal revenue that can be generated if player i invests w in working with $T_{i,j}$. We also write $T_{i,0}$ to be the tree T_i with all subtrees removed, i.e. the tree comprised of the singleton $\{i\}$. Now, suppose that we have already computed $v^*_{T_{i,j'}}(w)$ for all $j' = 0, \ldots, j-1$ and all $w = 0, \ldots, W_i$. In that case, we can compute $v^*_{T_{i,j}}(w)$ in time polynomial in W_M and n, as

$$v^*_{T_{i,j}}(w) = \max_{\substack{x+z=w \\ y \leq W_{c_j}}} \{v^*_{i,c_j}(x,y) + v^*_{T_{i,j-1}}(z) + v^*_{T_{c_j}}(W_{c_j} - y)\},$$

allowing us to find the value of $v^*_{T_{i,j}}(w)$ in time polynomial in W_M, and $v^*_{T_i}(0), \ldots, v^*_{T_i}(W_i)$ in time polynomial in W_M but linear in $|C_i|$. Going through all the nodes from the leaves to the root, we obtain that the value of the optimal coalition structure is simply the value $v^*_{T_r}(W_r)$. □

3.3 Computing Optimal Deviations

In Section 3.2 we identified three key conditions for the computational tractability of *finding an optimal coalition structure* in OCF games. Before we turn to computational aspects of *stability* in OCF games, let us study the problem of deciding the most that a set can obtain by deviating from a given outcome. Formally, we are interested in the following problem:

Definition 3.7 (ARBVAL). An instance of the ARBVAL problem is given by a $\langle \mathcal{G}, \mathcal{A}, S, (CS, \mathbf{x}), V \rangle$ tuple, where $\mathcal{G} = \langle N, \mathbf{W}, v \rangle$ is a discrete OCF game, $\mathcal{A}$ is some arbitration function, S is a subset of N, $(CS, \mathbf{x})$ is an outcome of $\mathcal{G}$, and V is some real value. It is a "yes" instance if $\mathcal{A}^*(CS, \mathbf{x}, S) \geq V$, and is a "no" instance otherwise.

We note that any hardness results obtained for OPTVAL are immediately inherited by ARBVAL, as the two problems are identical if one assumes that non-deviators behave according to the conservative arbitration function—i.e., give zero payoffs to deviators regardless of the nature of their deviation. However, ARBVAL is a considerably more complex problem than OPTVAL; in order to ensure that ARBVAL can be decided in polynomial time, one must make assumptions not only on the structure of the game $\mathcal{G}$, but also on the way that players react to deviation. In a sense, computational complexity can arise by making the payment structure of $\mathcal{A}$ sufficiently complex. This is shown in the following proposition.

Proposition 3.8. *If there exists a poly-time algorithm that can decide* ARBVAL, *when restricted to instances* $\langle \mathcal{G}, \mathcal{A}, S, (CS, \mathbf{x}), V \rangle$ *such that* $\mathcal{G} = \langle N, \mathbf{W}, v \rangle$ *is a 2 player game, then* $P = NP$.

Algorithm 2: Computing an optimal coalition structure in 2-OCF games with a tree interaction graph.

```
   // T is the tree interaction graph of G, and r ∈ N is an
      arbitrarily chosen root
1  Algorithm OptValTree(G = ⟨N, W, v⟩, T, r)
2  |  ℓ ← the number of children of r;
3  |  return OptValTreeSubroutine(G, T, r, r, W_r, ℓ);

   // Given a player i ∈ N, a discrete OCF game G with a tree
      interaction structure T rooted at r, and a value w ≤ W_i,
      the procedure outputs the maximal revenue that can be
      generated by T_i if i invests w in working with its first j
      children.
4  Procedure OptValTreeSubroutine(G = ⟨N, W, v⟩, T, r, i, w, j)
5  |  M ← OptVal(G, w · e^{i});
6  |  if j = 0 then
7  |  |  return M;
8  |  c_j ← the j-th child of i;
9  |  ℓ ← the number of c_j's children;
10 |  for x = 0, ..., w do
11 |  |  z ← w − x;
12 |  |  for y = 0, ..., W_{c_j} do
13 |  |  |  M' ← OptVal(G, x · e^{i} + y · e^{c_j}) +
         OptValTreeSubroutine(G, T, r, i, z, j − 1) +
         OptValTreeSubroutine(G, T, r, c_j, y, ℓ);
14 |  |  |  M ← max{M, M'};
15 |  return M;
```

Proof. We will show that if such an algorithm exists, it can be used to solve instances of SET-COVER [Garey and Johnson, 1979]. Recall that an instance of SET-COVER is given by a set of elements A, a collection of subsets $\mathcal{S} = \{S_1, \ldots, S_t\} \subseteq 2^A$ and $\ell \in \mathbb{N}$; it is a "yes"-instance if A can be covered by at most ℓ sets from $\mathcal{S}$.

Given an instance of SET-COVER, $\langle A, \mathcal{S}, \ell \rangle$ such that $|\mathcal{S}| = t$, consider a 2-player discrete OCF game where $W_1 = W_2 = t + 2$. We define v in the following manner. First, players get a payoff of 1 for each unit of resource devoted to working alone, i.e. $v_1(x) = v_2(x) = x$ for all $x \in \{0, \ldots, t+2\}$. We also set $v_{1,2}(1,1) = 2$, and $v_{1,2}(2,2) = 10(t+2)$. All other coalitions can have an arbitrary value (we assume with no loss of generality that it is 0).

We define an outcome $(CS, \mathbf{x})$ as follows. players 1 and 2 form t coalitions where each one of them devotes 1 unit of resources to working together, and an additional coalition where both invest 2 units; that is, $CS = (\mathbf{c}_1, \ldots, \mathbf{c}_t, \mathbf{d})$, such that $\mathbf{c}_j = (1, 1)$ for all $1 \leq j \leq t$, and $\mathbf{d} = (2, 2)$. We define $\mathbf{x} = (\mathbf{x}_1, \ldots, \mathbf{x}_t, \mathbf{y})$ as follows: $\mathbf{x}_j = (0, 2)$ is the payoff division from coalition $\mathbf{c}_j$, and $\mathbf{y} = (5(t+2), 5(t+2))$. In other words, we allocate the payoffs from $\mathbf{c}_1, \ldots, \mathbf{c}_t$ to player 2, and split the payoff from $\mathbf{d}$ equally.

We define the arbitration function $\mathcal{A}$ as follows. Given that player 1 wishes to deviate from $(CS, \mathbf{x})$, by withdrawing only from the coalitions $\mathbf{c}_{j_1}, \ldots, \mathbf{c}_{j_s}$, he receives no payoff from the coalitions $\mathbf{c}_1, \ldots, \mathbf{c}_t$, and will only get to keep his payoff from $\mathbf{d}$ if the collection $\{S_j \in \mathcal{S} \mid j \notin \{j_1, \ldots, j_s\}\}$ is a set cover of A. For any other input, we can define an arbitrary output for $\mathcal{A}$ (without loss of generality, let us assume that $\mathcal{A}$ behaves as the refined arbitration function on other inputs). Under this arbitration function, player 1 wants to withdraw as much resources as possible from $\mathbf{c}_1, \ldots, \mathbf{c}_t$, but do so in a manner that the coalitions he keeps intact correspond to a set cover of A.

We observe that $\mathcal{A}^*(CS, \mathbf{x}, \{1\}) \geq 5(t+2) + t - \ell$ if and only if $\langle A, \mathcal{S}, \ell \rangle$ is a "yes" instance of SET-COVER. First, if there is a subset $\mathcal{S}' \subseteq \mathcal{S}$ such that $|\mathcal{S}'| \leq \ell$, then by withdrawing from the coalitions corresponding to $\mathcal{S} \setminus \mathcal{S}'$, and allocating the withdrawn resources to working alone, player 1 ensures that he receives a payoff of at least $5(t+2) + t - \ell$; on the other hand, if there exists a set of coalitions $CS' = (\mathbf{c}_{j_1}, \ldots, \mathbf{c}_{j_s})$ such that withdrawing from CS' ensures that the payoff to player 1 is at least $5(t+2) + t - \ell$, then it must be that the set of coalitions that player 1 chose to withdraw from does not contain $\mathbf{d}$, and corresponds to a set $\mathcal{S}' \subseteq \mathcal{S}$ such that $\mathcal{S} \setminus \mathcal{S}'$ is a set cover of A. □

Proposition 3.8 is perhaps unsurprising: if the way non-deviators react to deviations is complex enough, then it would be computationally hard to decide whether deviation is worthwhile. However, we would like to point out the usefulness of this approach as a method of guaranteeing stability in OCF games. If we assume that players have limited computational power, then by making $\mathcal{A}$ complex enough, one can make an outcome $(CS, \mathbf{x})$ $\mathcal{A}$-stable in practice: even if a set S is able to deviate, it would be computationally infeasible for it to realize this. Thus, one is able to stabilize OCF games via *bureaucratic means*: making the rules that govern deviations so complex that it becomes computationally hard to decide whether changing the current state of affairs is profitable. Erecting computational barriers against undesirable behavior is common in other research fields, such as computational social choice.

Remark 3.9. We contrast Proposition 3.8 with Proposition 3.3: computing the most that a set can gain with a given set of resources is computationally much easier than deciding what is the most it stands to gain by deviating. This issue does not arise in classic cooperative games: a set assesses the desirability of deviation by considering the most it can make on its own, which only requires computing $u(S)$.

It seems that the hardness of deciding ARBVAL stems from the fact that the payoff from a coalition **c** to a deviating set S is determined by the way S affects other coalitions. In the reduction used in Proposition 3.8, the payoff to player 1 from the coalition **d** was determined by the deviation from other coalitions. In other words, the arbitration function determines the payoff to a deviating set S based on the global behavior of S.

This observation motivates the following definition.

Definition 3.10. An arbitration function $\mathcal{A}$ is *local* if the payoff from a coalition **c** depends only on the effect of the deviating set S on **c**, regardless of the input to $\mathcal{A}$. In other words, for any game $\mathcal{G} = \langle N, \mathbf{W}, v \rangle$, any outcome $(CS, \mathbf{x})$, any set $S \subseteq N$ and any deviation CS' of S from CS, the payoff to S from $\mathbf{c} \in CS \setminus CS|_S$ depends only on $\mathbf{d}(\mathbf{c})$, $\mathbf{x}(\mathbf{c})$, S, and $\mathcal{G}$.

The conservative, refined and optimistic arbitration functions are local: the payoff from the the conservative arbitration function is 0 for all inputs; the payment from the refined arbitration function is $x(S)$ if $\mathbf{d}(\mathbf{c}) = 0^n$ and is 0 otherwise, and the payment from the optimistic arbitration function is $\max\{v(\mathbf{c} - \mathbf{d}(\mathbf{c})) - \sum_{i \notin S} x(\mathbf{c})^i, 0\}$. In contrast, the arbitration function used the proof of Theorem 3.8 is non-local. Another example of a non-local arbitration function is the sensitive arbitration function, as the payoff to a set from a coalition **c** depends on which players in the support of **c** were hurt by the deviation of S from other coalitions.

When one is limited to the class of local arbitration functions, it is indeed possible to decide ARBVAL in time polynomial in $|CS|$ and $(W_M + 1)^{|S|}$, where S is the deviating set.

Theorem 3.11. ARBVAL *is decidable in time polynomial in* $|CS|$ *and* $(W_M + 1)^{|S|}$ *for all instances* $\langle \mathcal{G}, \mathcal{A}, S, (CS, \mathbf{x}), V \rangle$ *such that* $\mathcal{A}$ *is local.*

Proof. We show that Algorithm 3 computes $\mathcal{A}^*(CS, \mathbf{x}, S)$ in the required time, if $\mathcal{A}$ is local. We first observe that a coalition structure CS has at most $(W_M + 1)|S|$ coalitions that involve players in S. Given a coalition structure CS, let CS' be the set of coalitions that are supported by both S and $N \setminus S$; i.e. $CS' = \{\mathbf{c} \in CS \mid supp(\mathbf{c}) \cap S \neq \emptyset; supp(\mathbf{c}) \cap N \setminus S \neq \emptyset\}$. Now, suppose that players in S invest $\mathbf{s} \in \mathcal{W}(S)$ units of resources in partial coalitions among themselves —i.e. $\mathbf{w}(CS|_S) = \mathbf{s}$— and want to withdraw an additional $\mathbf{t} \in \mathcal{W}(S)$ from CS. They would get $v^*(\mathbf{s} + \mathbf{t})$ from working on their own, plus the most that S can get from the arbitration function, which depends on the coalitions affected by this deviation. Thus, in order to determine the most that S can get by deviating from $(CS, \mathbf{x})$, given that it must withdraw a total of **t** resources, we must determine how to best withdraw those **t** resources from CS'. We write $CS' = (\mathbf{c}_1, \ldots, \mathbf{c}_m)$.

Let us denote by $A(\mathbf{t}; \ell)$ the most that the arbitration function $\mathcal{A}$ will give S if they withdraw **t** resources from the first ℓ coalitions, where $1 \le \ell \le m$. We also write $\alpha_{\mathbf{c}}(\mathbf{t})$ to be the payoff to S from the coalition $\mathbf{c}$ if S withdraws **t** from $\mathbf{c}$. Since players may not withdraw more resources than they have invested in a coalition, we set $\alpha_{\mathbf{c}}(\mathbf{t}) = -\infty$ if **t** is greater than $\mathbf{c}$ in any coordinate. We note that $\alpha_{\mathbf{c}}(0^n)$ is the amount that S receives from $\mathbf{c}$ if it does not withdraw any resources from $\mathbf{c}$.

By definition, $A(\mathbf{t}; 1) = \alpha_{\mathbf{c}_1}(\mathbf{t})$, for other coalitions we have

$$A(\mathbf{t}; \ell) = \max\{A(\mathbf{y}; \ell - 1) + \alpha_{\mathbf{c}_\ell}(\mathbf{t} - \mathbf{y}) \mid 0^n \le \mathbf{y} \le \mathbf{t}\}.$$

This shows that we can compute $A(\mathbf{t}; m)$ in $\mathcal{O}(m(W_M + 1)^{|S|})$ steps. Finally, $\mathcal{A}^*(CS, \mathbf{x}, S)$ can be computed as $\max\{v^*(\mathbf{s} + \mathbf{t}) + A(\mathbf{t}; m) \mid 0^n \le \mathbf{t} \le \mathbf{W}|_S - \mathbf{s}\}$, which concludes the proof. $\square$

Now, as is the case for computing an optimal coalition structure for games where the interaction graph is a tree, computing the most that a set can get by deviating can be done in time polynomial in n and W_M if the arbitration function is local and the interaction graph is a tree. In fact, one can see Theorem 3.12 as an immediate corollary of Theorem 3.6.

Theorem 3.12. ARBVAL *is decidable in time polynomial in n and $(W_M + 1)$ for all instances $\langle \mathcal{G}, \mathcal{A}, S, (CS, \mathbf{x}), V \rangle$ such that $\mathcal{A}$ is local and such that $\mathcal{G}$ is a discrete 2-OCF game with a tree interaction structure.*

Proof. Again, we choose an arbitrary $r \in N$ to be the root of the interaction tree. Consider some player $i \in S$; i needs to decide how much weight to allocate to each non-S neighbor, given the amount that he gives to his parent. Let D_i be the non-S neighbors of i that i gets some payoff from interacting with under $(CS, \mathbf{x})$. Let us denote by $\alpha_i(w)$ the most that i can make if he devotes w to interacting with D_i. In order to compute $\alpha_i(w)$, we use dynamic programming. We fix an ordering of D_i and let $\alpha_i(w; j)$ be the most that player i can get by keeping a weight of w in the interaction with the first j players in D_i. We also denote by $\mathcal{A}_i^*(w; j)$ the payoff to player i if he keeps a weight of w in the interaction with the j-th player in D_i. By Theorem 3.11, $\mathcal{A}_i^*(w; j)$ can be computed in time polynomial in W_M. Further, we have $\alpha_i(w; 0) = 0$ for all w and

$$\alpha_i(w; j) = \max\{\alpha_i(x; j-1) + \mathcal{A}_i^*(w - x; j) \mid 0 \le x \le w\},$$

Set

$$\bar{v}_i^*(w) = \max\{v_i^*(x) + \alpha_i(w - x) \mid 0 \le x \le w\}.$$

We now replace player i's v_i^* by $\bar{v}_i^*$ for the purpose of computing $v^*(S)$, which is doable in polynomial time according to Theorem 3.6; this will give us the most that S can get by deviating under $\mathcal{A}$. $\square$

We observe that Theorem 3.12 holds even if the overall interaction graph is not a tree; it suffices that the deviating set S is an acyclic subgraph of the interaction graph. This is because in 2-OCF games, interactions can only be between pairs

Algorithm 3: Computing the most that a set can get by deviating, if $\mathcal{A}$ is local.

    // Computes the most that a set S can get by deviating from $(CS, \mathbf{x})$.
1. **Algorithm** ArbVal$(\mathcal{G} = \langle N, \mathbf{W}, v \rangle, \mathcal{A}, (CS, \mathbf{x}), S)$
2. $\mathbf{s} \leftarrow \mathbf{w}(CS|_S)$;
3.   let $CS' = (\mathbf{c}_1, \ldots, \mathbf{c}_m)$ be all coalitions $\mathbf{c}$ in CS such that $supp(\mathbf{c}) \cap S \neq \emptyset$ and $supp(\mathbf{c}) \cap (N \setminus S) \neq \emptyset$;
4. $M \leftarrow 0$;
5.   **for** $\mathbf{t} \leq \mathbf{W}|_S - \mathbf{s}$ **do**
6. $M' \leftarrow $ OptVal$(\mathcal{G}, \mathbf{s} + \mathbf{t}) + $ ArbPayoff$(\mathcal{G}, \mathcal{A}, (CS, \mathbf{x}), S, \mathbf{t}, m)$;
7. $M \leftarrow \max\{M, M'\}$;
8.   **return** M;

    // computes the most that the arbitration function will give a set S, if S wants to withdraw a total of $\mathbf{t}$ resources from the first ℓ coalitions in CS'.
9. **Procedure** ArbPayoff$(\mathcal{G}, \mathcal{A}, (CS, \mathbf{x}), S, \mathbf{t}, \ell)$
    // If $\ell = 0$ then S withdraws no resources from any coalition.
10.   **if** $\ell = 0$ **then**
11.   **return** 0;
12. $M \leftarrow 0$;
13.   **for** $0^n \leq \mathbf{y} \leq \mathbf{t}$ **do**
        // $\alpha_{\mathbf{c}}(\mathbf{q})$ is what S gets according to $\mathcal{A}$, if it withdraws $\mathbf{q}$ resources from the coalition $\mathbf{c}$; if $\mathbf{q}$ is not coordinate-wise smaller than $\mathbf{c}$, then $\alpha_{\mathbf{c}}(\mathbf{q}) = -\infty$
14. $M' \leftarrow \alpha_{\mathbf{c}_\ell}(\mathbf{t} - \mathbf{y}) + $ ArbPayoff$(\mathcal{G}, \mathcal{A}, (CS, \mathbf{x}), S, \mathbf{y}, \ell - 1)$;
15. $M \leftarrow \max\{M, M'\}$;
16.   **return** M;

of players; thus, if a player $i \in S$ decides to withdraw resources from a coalition **c**, that coalition can contain at most one other player j. Therefore, no other player in S is concerned with **c**, i.e. players in S need to keep individual track of how much they withdraw from coalitions supported by $N \setminus S$, which significantly simplifies the computational process of deciding where to withdraw resources from.

3.4 Computing $\mathcal{A}$-Stable Outcomes

Having provided efficient procedures for computing optimal deviations and coalition structures in discrete 2-OCF games with tree interaction structures, we are ready to analyze the computational complexity of stability in this class of games. Recall that an OCF game is $\mathcal{A}$-stable if there exists some outcome $(CS, \mathbf{x})$ such that no subset of N can profitably deviate from $(CS, \mathbf{x})$, i.e. if for all $S \subseteq N$ we have that

$$p_S(CS, \mathbf{x}) \geq \mathcal{A}^*(CS, \mathbf{x}, S).$$

In this context, we are first interested in the following problem.

Definition 3.13 (CHECKCORE). An instance of CHECKCORE is given by a discrete OCF game $\mathcal{G} = \langle N, \mathbf{W}, v \rangle$, an arbitration function $\mathcal{A}$, and an outcome $(CS, \mathbf{x})$; it is a "yes" instance if $(CS, \mathbf{x})$ is in the $\mathcal{A}$-core of $\mathcal{G}$, and a "no" instance otherwise.

We are now ready to present an algorithm for checking whether a given outcome is in the $\mathcal{A}$-core. This problem is closely related to that of computing $\mathcal{A}^*$: an outcome $(CS, \mathbf{x})$ is in the $\mathcal{A}$-core if and only if the *excess* $e(CS, \mathbf{x}, S) = \mathcal{A}^*(CS, \mathbf{x}, S) - p_S(CS, \mathbf{x})$ is non-positive for all coalitions $S \subseteq N$. Thus, we need to check whether there exists a subset $S \subseteq N$ with $e(CS, \mathbf{x}, S) > 0$. Note that it suffices to limit attention to subsets of N that form connected subgraphs of the interaction graph of the game: if $e(CS, \mathbf{x}, S) > 0$ and S is not connected, then some connected component S' of S also satisfies $e(CS, \mathbf{x}, S') > 0$.

Theorem 3.14. *If $\mathcal{G}$ is a 2-OCF game with a tree interaction structure, and $\mathcal{A}$ is local, then* CHECKCORE *is decidable in time polynomial in n and $(W_M + 1)$ for any instance $\langle \mathcal{G}, \mathcal{A}, (CS, \mathbf{x}) \rangle$.*

Proof. Fixing an outcome $(CS, \mathbf{x})$, let us write $p_i = p_i(CS, \mathbf{x})$ for all $i \in N$.

Again, we pick an arbitrary $r \in N$ as a root. We say that $S \subseteq N$ is *rooted* at $i \in N$ if $i \in S$ and the members of S form a subtree of T_i - the tree rooted in i. We observe that every set $S \subseteq N$ is rooted at a unique $i \in N$. Given a vertex i, let E_i denote the maximum excess of a set rooted at i, that is:

$$E_i = \max \{e(CS, \mathbf{x}, S) \mid S \text{ is rooted in } T_i\}.$$

Clearly, $(CS, \mathbf{x})$ is not $\mathcal{A}$-stable if and only if $E_i > 0$ for some $i \in N$. It remains to show that all E_i can be computed in time polynomial in n and W_M. As before, we proceed from the leaves to the root, and terminate (and report that $(CS, \mathbf{x})$ is not $\mathcal{A}$-stable) if we discover a vertex i with $E_i > 0$. If $E_i \leq 0$ for all $i \in N$, we report that $(CS, \mathbf{x})$ is $\mathcal{A}$-stable.

Given two players $i, j \in N$, let $w_{i,j}$ denote the total weight that i assigns to interacting with j under CS, i.e.

$$w_{i,j} = \sum_{\mathbf{c}: supp(\mathbf{c}) = \{i,j\}} c_i.$$

We begin by defining two auxiliary values. First, given a neighbor j of i, we define $\alpha_{i,j}(w)$ to be the most that $\mathcal{A}$ will give i if he keeps a total weight of $w \leq w_{i,j}$ in the coalitions that he formed with j in $(CS, \mathbf{x})$; by Theorem 3.11, $\alpha_{i,j}(w)$ is computable in time polynomial in $(W_M + 1)$. Second, we define $D_i(w)$ to be the maximum excess of a subset rooted at i if i were to contribute w to T_i and nothing to his parent $p(i)$. In this notation,

$$E_i = \max\{D_i(w) + \alpha_{i,p(i)}(y) \mid w + y = w_i, y \leq w_{i,p(i)}\};$$

the condition $y \leq w_{i,p(i)}$ ensures that i does not give additional resources to working with $p(i)$; that is, $p(i)$ is not among the deviators. It remains to show how to compute $D_i(w)$ in time $\mathrm{poly}(n, W_M)$ for all $i \in N$ and all $y \in \{0, \ldots, w_{i,p(i)}\}$.

Consider a player i with children $C_i = \{i_1, \ldots, i_\ell\}$, and suppose that we have computed $D_{i_j}(z)$ for each $i_j \in C_i$ and each z such that $w_{i_j} - w_{i_j,i} \leq z \leq w_{i_j}$ (this encompasses the possibility that i is a leaf, as $C_i = \emptyset$ in that case). For $j = 0, \ldots, \ell$, let $T_{i,j}$ be the tree obtained from T_i by removing subtrees rooted at $i_{j+1}, \ldots, i_\ell$. Let $D_i(w; j)$ be the maximum excess of a set rooted at i that is fully contained in $T_{i,j}$, assuming that i contributes w to $T_{i,j}$ and nothing to his parent or his children $i_{j+1}, \ldots, i_\ell$; we have $D_i(w) = D_i(w; \ell)$. We will compute $D_i(w; j)$ by induction on j.

We have $D_i(w; 0) = v_i^*(w) - p_i$ for all $w = w_i - w_{i,p(i)}, \ldots, w_i$. Now, consider $j > 0$. Player i can either include i_j in the deviating set or deviate (partially or fully) from the coalitions that it forms with i_j in $(CS, \mathbf{x})$. Thus, $D_i(w; j) = \max\{D_1, D_2\}$, where

$$D_1 = \max_{\substack{y=0,\ldots,w \\ z=0,\ldots,w_{i_j}}} \{D_i(y; j-1) + v_{i,i_j}^*(w - y, z) + D_{i_j}(w_{i_j} - z)\}.$$

and

$$D_2 = \max_{z=0,\ldots,w_{i,i_j}} \{D_i(w - z; j-1) + \alpha_{i,i_j}(z)\}.$$

Since both quantities D_1 and D_2 can be computed in time polynomial in W_M, we can efficiently compute $D_i(w; j)$, and hence also $D_i(w)$ and E_i. □

Not only can the algorithm described in Theorem 3.14 able to decide CHECKCORE in time polynomial in n and W_M, it can also be used to decide the following, closely related problem. We are not only interested in deciding whether a given outcome $(CS, \mathbf{x})$ is $\mathcal{A}$-stable: we are also interested in deciding whether a given coalition structure CS can be stabilized; i.e. whether there exists some division of payoffs among players in such a way that the resulting outcome is $\mathcal{A}$-stable. This is formalized in the following definition.

Definition 3.15. An instance of IS-STABLE is given by a discrete OCF game $\mathcal{G} = \langle N, \mathbf{W}, v \rangle$, an arbitration function $\mathcal{A}$ and a coalition structure CS.

Given a coalition structure $CS = (\mathbf{c}_1, \ldots, \mathbf{c}_m)$, an instance of IS-STABLE $\langle \mathcal{G}, \mathcal{A}, CS \rangle$ is a "yes" instance if and only if there exists a payoff division $\mathbf{x} \in I(CS)$, $\mathbf{x} = (\mathbf{x}_1, \ldots, \mathbf{x}_m)$, that satisfies the following system of constraints.

$$\sum_{i \in supp(\mathbf{c}_j)} x_{ji} = v(\mathbf{c}_j) \qquad \forall j \in \{1, \ldots, m\} \qquad (3.1)$$

$$\sum_{i \in S} \sum_{j=1}^{m} x_{ji} \geq \mathcal{A}^*(CS, \mathbf{x}, S) \qquad \forall S \subseteq N$$

$$x_j^i \geq 0 \quad \forall j \in \{1, \ldots, m\}; \forall i \in supp(\mathbf{c}_j)$$

The first set of constraints —also called *efficiency* constraints— ensures that $\mathbf{x}$ is indeed a valid imputation, while the second set of constraints —also called *stability* constraints— ensures that $(CS, \mathbf{x})$ is in the $\mathcal{A}$-core of $\mathcal{G}$. The number of constraints in (3.1) is exponential in n; moreover, there is no guarantee that $\mathcal{A}^*(CS, \mathbf{x}, S)$ is a value that is linear in $\mathbf{x}$. However, if the set of constraints described in (3.1) is linear in $\mathbf{x}$, then one can use a simple modification of the algorithm described in Theorem 3.14 as a *separation oracle*; given a linear program with an arbitrary number of constraints, a separation oracle is an algorithm whose input is a candidate point, that can decide in polynomial time whether the point satisfies all constraints, and if not, can output a violated constraint. Given such an algorithm, one can decide if there exists a point that satisfies all constraints.

We note that the set of constraints (3.1) is linear for the conservative, refined, and optimistic arbitration functions. Thus, we obtain the following corollary.

Corollary 3.16. IS-STABLE *is decidable in time polynomial in n and W_M for all instances $\langle \mathcal{G}, \mathcal{A}, CS \rangle$ such that $\mathcal{G}$ is a 2-OCF game with a tree interaction structure, and $\mathcal{A}$ is the conservative, refined or optimistic arbitration function.*

Indeed, Corollary 3.16 holds for *any* local arbitration function for which the set of constraints (3.1) is linear.

3.5 Beyond Tree Interactions

In previous sections, we have shown that if the game $\mathcal{G}$ is a discrete 2-OCF game with a tree interaction structure, then most relevant stability notions can be computed in time polynomial in n, the number of players, and $(W_M + 1)$, the maximal weight of any player. We now show how to extend our algorithms to 2-OCF games that have an interaction graph that is not a tree.

If $\mathcal{G}$ is a 2-OCF game, then its interaction graph is a simply a graph with either simple edges or self edges. In this section, we show how our algorithms and their complexity can be parameterized by the *treewidth* [Robertson and Seymour, 1984] of the game's interaction graph. The algorithms we describe assume that the interaction graph is connected; however, all our results hold even if the interaction graph is not connected, by simply applying our methods to each of the connected components separately. Let us begin by defining the notion of treewidth. Given a graph $\Gamma = \langle N, E \rangle$, a *tree decomposition* of Γ is a tree $\mathcal{T}$ whose vertices are subsets of N (we write $V(\mathcal{T})$ to denote the vertices of $\mathcal{T}$ and $E(\mathcal{T})$ to denote its edges), and which satisfies the following three conditions

1. If $e \in E$ then there is some vertex $S \in V(\mathcal{T})$ such that $e \subseteq S$.

2. Given any two vertices $S, S' \in V(\mathcal{T})$ such that there is some $i \in N$ that is in $S \cap S'$, i appears in every vertex on the path between S and S'.

Given a tree decomposition $\mathcal{T}$ of Γ, let us write $\text{width}(\mathcal{T})$ to be $\max\{|S| \mid S \in V(\mathcal{T})\} - 1$; we define the *treewidth* of a graph Γ to be

$$\text{tw}(\Gamma) \stackrel{\text{def}}{=} \min\{\text{width}(\mathcal{T}) \mid \mathcal{T} \text{ is a tree decomposition of } \Gamma\}.$$

We note that the -1 is simply a normalization factor, which ensures that the treewith of trees is 1; in fact, it is well-known that a graph is a tree if and only if its treewidth is 1. Given a tree decomposition $\mathcal{T}$, we write $N(\mathcal{T}) \stackrel{\text{def}}{=} \bigcup_{S \in V(\mathcal{T})} S$; i.e. $N(\mathcal{T})$ is the set of all players that are in the nodes of $\mathcal{T}$.

Treewidth is often used as a parameter in the parameterized complexity analysis of graph related combinatorial problems; Courcelle's theorem [Courcelle, 1990] states that any graph property that can be stated using a fairly standard set of operators (monadic second order logic) is fixed parameter tractable, with the treewidth of the graph being the parameter. Treewidth has also been used in the study of cooperative games, both for studying the computational complexity of finding solution concepts [Greco et al., 2011], and in studying their structure [Meir et al., 2013].

We now generalize Theorems 3.6, 3.12 and 3.14 for 2-OCF games whose interaction graphs have a treewidth of k. We note that deciding whether an interaction graph has a treewidth of k (and finding a tree decomposition of Γ whose width is at most k) can be done in time polynomial in the number of nodes in Γ and exponential in k. Therefore, the computational complexity of the results that now follow does not increase if one first decides whether $\mathcal{G}$ has an interaction graph of width k. We overload notation and write $\text{tw}(\mathcal{G})$ to be the treewidth of the interaction graph of $\mathcal{G}$, where $\mathcal{G}$ is a discrete 2-OCF game.

Theorem 3.17. OPTVAL *is decidable in time polynomial in n and $(W_M + 1)^{\text{tw}(\mathcal{G})+1}$ for all instances $\langle \mathcal{G}, \mathbf{c}, V \rangle$ such that $\mathcal{G}$ is a 2-OCF game.*

Proof. We again show how to compute an optimal coalition structure when all players invest all their resources; the reduction to a general coalition $\mathbf{c}$ is trivial. Let $\mathcal{T}$ be a tree decomposition of the interaction graph of $\mathcal{G}$ such that $\text{width}(\mathcal{T}) = k$. Let us choose some $R \in V(\mathcal{T})$ to be the root of $\mathcal{T}$; for any $X \in V(\mathcal{T})$, let us write $\mathcal{T}_X$ the subtree rooted in the vertex X, $p(X)$ to the parent of X in $\mathcal{T}$, and C_X to be the children of X in $\mathcal{T}$. Intuitively, in order to compute an optimal coalition structure, players in $X \cap p(X)$ need to decide how much to allocate to their own subtree $\mathcal{T}_X$, and how much to allocate to working with their parent. Let us write $\text{opt}(\mathcal{T}_X(\mathbf{q}))$ to be the value of an optimal coalition structure over the nodes in $\mathcal{T}_X$, but with the players in $X \cap p(X)$ investing only $\mathbf{q}$ in working with $\mathcal{T}_X$. We observe

that
$$\mathrm{opt}(\mathcal{T}_X(\mathbf{q})) = \max\left\{v^*(\mathbf{y} + \sum_{Y \in C_X} \mathbf{x}_Y) + \sum_{Y \in C_Y} \mathrm{opt}(\mathcal{T}_Y(\mathbf{z}_Y))\right\},$$

where $\mathbf{z}_Y$ is the amount that the set $X \cap Y$ devotes to working with $\mathcal{T}_Y$, and $\mathbf{x}_Y$ is what is allocated to working with X; $\mathbf{y}$ is the vector of resources of $X \setminus \bigcup_{Y \in C_Y} Y$, assuming that those members of $X \cap p(X)$ contribute according to $\mathbf{q}$, i.e. $\mathbf{y} = \min\{\mathbf{q}, \mathbf{W}|_{X \setminus \bigcup_{Y \in C_Y} Y}\}$. Thus, it must hold that $\mathbf{y} + \sum_{Y \in C_Y} \mathbf{x}_Y + \mathbf{z}_Y = \min\{\mathbf{q}, \mathbf{W}^X\}$, and $\mathbf{x}_Y + \mathbf{z}_Y \leq \min\{\mathbf{q}, W|_{X \cap Y}\}$ for all $Y \in C_X$.

Taking a similar approach to that used in Theorem 3.6, we employ dynamic programming in order to compute $\mathrm{opt}(\mathcal{T}_X(\mathbf{q}))$. We write $\mathrm{opt}(\mathcal{T}_X(\mathbf{q}; j))$ to be the most that $\mathcal{T}_X$ can make if $X \cap p(X)$ allocates $\mathbf{q}$ to working with $\mathcal{T}_X$, and only the first j children are considered, where C_X is set to be $\{Y_1, \ldots, Y_m\}$. $\mathrm{opt}(\mathcal{T}_X(\mathbf{q}; 0))$ is simply $v^*(\min\{\mathbf{W}|_X, \mathbf{q}\})$, and for all $j \geq 1$:

$$\mathrm{opt}(\mathcal{T}_X(\mathbf{q}; j)) = \max\left\{\mathrm{opt}(\mathcal{T}_X(\mathbf{q} - \mathbf{z}; j - 1)) + \mathrm{opt}(\mathcal{T}_{Y_j}(\mathbf{z})) \mid \mathbf{z} \leq \min\{\mathbf{q}, \mathbf{W}^{X \cap Y_j}\}\right\}.$$

To conclude, assuming we have computed $\mathrm{opt}(\mathcal{T}_Y(\mathbf{z}))$ for all $Y \in C_X$ and all $\mathbf{z}$, we can compute $\mathrm{opt}(\mathcal{T}_X(\mathbf{q}))$ in time polynomial in $(W_M + 1)^{\mathrm{tw}(\mathcal{G})+1}$ and linear in $|C_X|$, which implies that the total running time of the dynamic program is polynomial in $(W_M + 1)^{\mathrm{tw}(\mathcal{G})+1}$ and linear in n. □

A similar approach can be used in order to compute the most that a set can get by deviating from an arbitrary graph. The same key observation used in Theorem 3.12 is made here: in order to compute the most that a set S can get by deviating, we first replace $v_i^*(w)$ with $\bar{v}_i^*(w)$, where $\bar{v}_i^*(w) = \max\{\alpha_i(w - x) + v_i^*(x) \mid 0 \leq x \leq w\}$, and $\alpha_i(w)$ is the most that i can get from the arbitration function if it leaves a total of w of its resources with non-deviators. Having replaced v_i^* with $\bar{v}_i^*$ we run the algorithm described in Theorem 3.17 to obtain the following:

Theorem 3.18. ARBVAL *is decidable in time polynomial in n and $(W_M + 1)^{\mathrm{tw}(\mathcal{G})+1}$ for all instances $\langle \mathcal{G}, \mathcal{A}, \mathbf{c}, V \rangle$ such that $\mathcal{A}$ is local and $\mathcal{G}$ is a 2-OCF game.*

Finally, we provide an algorithm for deciding instances of CHECKCORE that runs in time polynomial in n and $(W_M + 1)^{\mathrm{tw}(\mathcal{G})+1}$.

Theorem 3.19. *An instance of* CHECKCORE *is decidable in time polynomial in n and $(W_M + 1)^{\mathrm{tw}(\mathcal{G})+1}$ for all instances $\langle \mathcal{G}, \mathcal{A}, (CS, \mathbf{x}) \rangle$ such that $\mathcal{A}$ is local and $\mathcal{G}$ is a 2-OCF game.*

Proof. Given an outcome $(CS, \mathbf{x})$, our goal is to find a subset $S \subseteq N$ such that $e(CS, \mathbf{x}, S) < 0$ if such a subset exists. Let $\mathcal{T}$ be the tree decomposition of the interaction graph of $\mathcal{G}$, and we again choose some $R \in V(\mathcal{T})$ to be the root of $\mathcal{T}$. Take some $S \subseteq X$; let us denote by $\mathcal{T}_S$ the subtree rooted in X, but with the members of $X \setminus S$ removed from all the nodes in $\mathcal{T}_X$. We say that a subset T of N is rooted in $\mathcal{T}_S$ if $T \subseteq \mathrm{N}(\mathcal{T}_S)$ and $S \subseteq T$; that is, T is rooted in $\mathcal{T}_S$ only if it contains S, as well as being contained in $\mathrm{N}(\mathcal{T}_S)$. We write $E_S(\mathbf{q})$ to be the excess of the unhappiest subset rooted in S, assuming that S devotes only $\mathbf{q} \leq \mathbf{W}^S$ to interacting with $\mathcal{T}_S$.

Now, suppose that we have already computed $E_T(\mathbf{z})$ for all $T \subseteq Y$ where Y is a child of X and for all $\mathbf{z} \leq \mathbf{W}^Y$. Let us set $C_S = \{Y_1,\ldots,Y_m\}$, where $C_S = \{Y \setminus (N \setminus S) \mid Y \in C_X\}$; we write $E_S(\mathbf{q}; j)$ to be the maximal excess achievable by any subset rooted in $\mathcal{T}_S$, with all resources allocated to the first j children of S, and assuming that S allocates $\mathbf{q}$ resources to working with $\mathcal{T}_S$. Therefore, $E_S(\mathbf{q}; 0)$ is $e(CS, \mathbf{x}, S, \mathbf{q}) = \mathcal{A}^*(CS, \mathbf{x}, S, \mathbf{q}) - p_S(CS, \mathbf{x})$, where $\mathcal{A}^*(CS, \mathbf{x}, S, \mathbf{q})$ is the most that S can get by deviating when it has only $\mathbf{q}$ resources to allocate to working with non-deviators and optimize its own payoffs.

Now, when choosing how to deviate with the j-th child, S needs to decide how much of its resources to allocate to Y_j, and which subset of Y_j to join into the deviation. It has already joined all members of $S \cap Y_j$, but it now needs to choose an additional subset $T \subseteq Y_j \setminus S$ to bring into the deviation, and demand resources from it in an optimal manner; in other words,

$$E_S(\mathbf{q}; j) = \max \left\{ E_S(\mathbf{q}'; j-1) + v^*(\mathbf{q} - \mathbf{q}' + \mathbf{r}) + E_T(\mathbf{W}^T - \mathbf{r}) \right\},$$

where the maximization is over all $\mathbf{q}' \leq \mathbf{q}$, all $T \subseteq Y_j$ and all $\mathbf{r} \leq \mathbf{W}^T$.

Thus, we can compute $E_S(\mathbf{q})$ in time polynomial in $|C_X|$ and $(2(W_M + 1))^{\text{tw}(\mathcal{G})}$, and therefore decide CHECKCORE in polynomial time as well. □

We end this section with two concluding remarks. First, as was the case for trees, the algorithm described in Theorem 3.19 can be used as a separation oracle for deciding IS-STABLE in time polynomial in $(W_M + 1)^{\text{tw}(\mathcal{G})+1}$ and n for general 2-OCF games.

Second, we would like to mention that our theorems imply that when W_M is a constant independent of n, OPTVAL, ARBVAL, CHECKCORE and IS-STABLE are fixed-parameter tractable where $\text{tw}(\mathcal{G})$ is the parameter, and $\mathcal{G}$ is a 2-OCF game. Moreover, when $\mathcal{G}$ is a k-OCF game, rather than a 2-OCF game, we have that OPTVAL is fixed-parameter tractable when W_M is a constant.

3.6 Linear Bottleneck Games and the Optimistic Core

For the final section of this chapter, let us return to the standard (non-discrete) OCF setting. The main objective of this section is to describe a class of cooperative games with overlapping coalitions that has a non-empty optimistic core, and instances of OPTVAL, ARBVAL, and IS-STABLE are decidable in polynomial time when restricted to this class. Our class of OCF games is motivated by fractional combinatorial optimization scenarios. In previous sections, we make no assumptions on the structure of the characteristic function, but rather use underlying player interaction to facilitate poly-time computation. In what follows, we do not make any assumptions on player interactions, but rather restrict our attention to a family of characteristic functions. This approach leads to the strong poly-time results presented here.

We begin by recalling the notion of stability under the optimistic arbitration function, which will be notion of stability we study in this section. Given the optimistic arbitration function, denoted $\mathcal{A}_o$, an $\mathcal{A}_o$-profitable deviation of a set $S \subseteq N$ from an outcome $(CS, \mathbf{x})$ can be described by

(a) the list of coalitions $CS|_S \subseteq CS' \subseteq CS$ that S fully withdraws resources from. These are coalitions that S does not wish to retain payoffs from, thus it fully withdraws its resources from them, and utilizes those resources to maximize its own profits.

(b) the partial deviation of S CS'' from $CS \setminus CS'$, i.e. the amount of resources each $i \in S$ withdraws from each coalition in $CS \setminus CS'$. The coalitions in CS' are those that S does wish to retain interactions with, and is thus willing to maintain the payoffs to $N \setminus S$ in those coalitions, effectively assuming the marginal cost of its deviation from those coalitions.

S is then allowed to use the resources which it has withdrawn according to CS' and CS'' in order to maximize its own profits, while absorbing the damage it has caused $CS \setminus CS'$ by withdrawing CS''. We define a large class of OCF games that is motivated by combinatorial optimization and resource allocation scenarios, and prove that these games always have a non-empty optimistic core. Moreover, we show that for games in this class an optimal coalition structure can be found using linear programming, and the dual LP solution can be used to find an imputation in the optimistic core. Our results in this section build on prior work on classic cooperative game theory, where primal-dual methods have been used to derive explicit payoff divisions that guarantee core stability [Deng et al., 1999, Jain and Mahdian, 2007, Markakis and Saberi, 2005]; indeed, one can view our results as stating that not only are the games described in these works stable against deviations in the classic cooperative sense (i.e., have a non-empty conservative core), they are also resistant to deviations when much more lenient player behavior is assumed. First, let us define the class of games we are interested in.

Definition 3.20. A *Linear Bottleneck Game* $\mathcal{G} = (N, \mathbf{W}, \mathcal{T})$ is given by a set of players $N = \{1, \ldots, n\}$, a list $\mathbf{W} = (W_1, \ldots, W_n)$ of players' *weights*, and a list of *tasks* $\mathcal{T} = (T_1, \ldots, T_m)$, where each task T_j is associated with a set of players $A_j \subseteq N$ who are needed to complete it, as well as a *value* $\pi_j \in \mathbb{R}_+$. We assume that $A_j \neq A_{j'}$ for $j \neq j'$, and for each $i \in N$ there is a task $T_k \in \mathcal{T}$ with $A_k = \{i\}$. The characteristic function of this game is defined as follows: given a partial coalition $\mathbf{c} \in [0, 1]^n$, we set

$$v(\mathbf{c}) = \begin{cases} \pi_j \cdot \min_{i \in A_j} c_i W_i & \text{if } supp(\mathbf{c}) = A_j \text{ for some } j \in [m] \\ 0 & \text{otherwise.} \end{cases}$$

These games are linear in the sense that the payoff earned by a partial coalition scales linearly with the smallest contribution to this coalition; the smallest contribution is the "bottleneck" contribution, since the contribution of no member but the smallest member affects the value of the coalition. The assumption that $A_j \neq A_{j'}$ for $j \neq j'$ ensures that the characteristic function is well-defined; that is, each task is associated with a unique set of players that can complete it. Finally, since each player can work on his own (possibly earning a payoff of 0), all resources are used. This assumption will be useful when proving our results, since it allows us to invest unused player resources in dummy tasks.

3.6.1 Some Examples

LBGs can be used to describe a variety of settings; a more complete overview of their descriptive power can be seen in Deng et al. [1999]. However, for the sake of completeness, we provide three examples below. First, LBGs can describe *multicommodity flow games* [Vazirani, 2001, Markakis and Saberi, 2005]. Briefly, in multicommodity flow games, pairs of vertices in a network want to send and receive flow, which has to be transmitted by edges of the network. This setting can be modeled by a linear bottleneck game, where both vertices and edges are players. The weight of an edge player is the capacity of his edge, while the weight of a vertex player is the amount of commodity he possesses.

More formally, there are two types of players in the multicommodity flow game: *suppliers*, denoted N_s, and *distributors*, denoted N_d. Given a directed graph Γ with an edge set $E(\Gamma)$ and a node set $V(\Gamma)$, each supplier $i \in N_s$ controls a pair of nodes (s_i, t_i), and has certain amount W_i of a *commodity*, with a per-unit price of π_i. Each distributor $j \in N_d$ controls an edge e_j, such that $\{e_j\}_{j \in N_d} = E(\Gamma)$; each edge has a weight, or capacity, $w(e)$. A task in this game is to transfer a certain amount of a commodity owned by i from s_i to t_i. In this setting, each path from s_i to t_i is a task, with its associated set being the distributors on the path, and the supplier controlling (s_i, t_i); the value of a coalition is the amount of the commodity supplied by i that the path transfers, times the per-unit value of that commodity. We note that in this description, the number of possible tasks may be exponential in the number of players, however, as Markakis and Saberi [2005] show, there do exist other, more succinct, ways of describing the problem that result in the same solution, and to which our techniques can be applied. We do maintain the current description, as it does highlight the fact that multicommodity flow games are indeed linear bottleneck games.

Another example of a linear bottleneck game occurs in network routing settings. Consider again a directed graph Γ with an edge set $E(\Gamma)$ and a node set $V(\Gamma)$. Here, players are nodes, and each node i has a weight W_i (this can be thought of as processing power, or amount of memory). Now, the tasks in this setting are to transfer data from certain source nodes to certain target nodes, with each such (s_j, t_j) pair associated with a per-unit payoff π_j. Unlike multicommodity flow games, the amount of data to be transferred is unlimited, with the only limitation on transfer power stemming from players' own capacity constraints.

Finally, consider a bipartite graph with the node sets A, B, such that $A \cap B = \emptyset$, and with no edges among the members of A or the members of B. Players are nodes, and for every $a \in A$ and $b \in B$, the edge $e = \{a, b\}$ has a value π_e. Each player $i \in A \cup B$ has a weight W_i. The tasks here are the edges, with the member nodes of each edge being the set required to complete the task, and the payoff being π_e. This setting can be thought of as a generalized fractional weighted matching game, or, alternatively, as a trading market, where A is a set of sellers and B is a set of buyers. Having $a \in A$ and $b \in B$ form a coalition means that b

agrees to buy from a for the set price of π_e per unit. The total value of the coalition structure can be thought of as the total volume of exchanges made in the trading market.

3.6.2 Computing Stable Outcomes in LBGs

Before we proceed, let us make some simple observations on the structure of optimal coalition structures in LBGs.

Lemma 3.21. *Given an LBG $\mathcal{G} = \langle N, \mathbf{W}, \mathcal{T} \rangle$, there is some optimal coalition structure CS such that*

(a) for all $\mathbf{c} \in CS$ we have $c_i W_i = c_k W_k$ for all $i, k \in supp(\mathbf{c})$.

(b) $w_i(CS) = 1$ for all $i \in N$.

(c) each A_j forms at most one coalition in CS.

Proof. First, given an optimal coalition structure CS for a linear bottleneck game, we can assume without loss of generality that for every $\mathbf{c}$ in CS and every $i, k \in A_j$ we have $c_i W_i = c_k W_k$: investing more weight than one's team members does not increase the payoff from the task, so a player might as well use this weight to work alone. Second, since we assume that there is a task that a player can complete alone, the value of a coalition structure can only increase when players invest any unused weight in working alone. Finally, it can be assumed that CS contains at most one coalition $\mathbf{c}$ with $supp(\mathbf{c}) = A_j$ for each $j = 1, \ldots, m$: if $supp(\mathbf{c}) = supp(\mathbf{d}) = A_j$, then $v(\mathbf{c} + \mathbf{d}) \geq v(\mathbf{c}) + v(\mathbf{d})$, so two coalitions with the same support can be merged. This implies that we can assume that in an optimal coalition structure each A_j forms at most one coalition $\mathbf{c}_j$. □

Lemma 3.21 implies that an optimal coalition structure can be described by a list $C_1, \ldots C_m$, indicating how much weight is allocated to each task.

Therefore, according to Lemma 3.21, the following a linear program finds an optimal coalition structure for an LBG $\mathcal{G} = \langle N, \mathbf{W}, \mathcal{T} \rangle$:

$$\begin{aligned} \max: \quad & \sum_{j=1}^{m} C_j \pi_j \\ \text{s.t.} \quad & \sum_{j: i \in A_j} C_j \leq W_i \quad \forall i \in N \\ & C_j \geq 0 \quad \forall j \in [m] \end{aligned} \quad (3.2)$$

The dual of LP (3.2) is

$$\begin{aligned} \min: \quad & \sum_{i=1}^{n} W_i \gamma_i \\ \text{s.t.} \quad & \sum_{i \in A_j} \gamma_i \geq \pi_j \quad \forall j \in [m] \\ & \gamma_i \geq 0 \quad \forall i \in N \end{aligned} \quad (3.3)$$

Let $\widehat{C_1}, \ldots, \widehat{C_m}$ and $\widehat{\gamma_1}, \ldots, \widehat{\gamma_n}$ be optimal solutions to (3.2) and (3.3) respectively. Let CS be the coalition structure that corresponds to $\widehat{c}_1, \ldots, \widehat{c}_m$. We construct a payoff vector $\mathbf{x}$ for CS as follows: for every $j = 1, \ldots, m$ we set $x_{ji} = \widehat{\gamma_i}\widehat{C_j}$, if $i \in A_j$, and $x^{ji} = 0$ otherwise. In words, each player i has some "bargaining power" $\widehat{\gamma_i}$, and is paid for each task he works on in proportion to his bargaining power. Note that both CS and x can be computed in time polynomial in n and $|\mathcal{T}|$. We will now show that x is an imputation for CS, and, moreover, $(CS, \mathbf{x})$ is in the optimistic core.

Theorem 3.22. *Let $\mathcal{G} = \langle N, \mathbf{W}, \mathcal{T} \rangle$ be a linear bottleneck game, and let CS and x be the coalition structure and the payoff vector constructed above. Then $\mathbf{x} \in I(CS)$ and $(CS, \mathbf{x})$ is in the optimistic core of $\mathcal{G}$.*

Proof. First, we argue that $\mathbf{x} \in I(CS)$. To see that x satisfies coalitional efficiency, note that the sum of payoffs from task T_j is

$$\sum_{i \in A_j} x_{ji} = \sum_{i \in A_j} \widehat{\gamma_i} \widehat{C_j}$$
$$= \widehat{C_j} \sum_{i \in A_j} \widehat{\gamma_i}$$

As $\widehat{\gamma_1}, \ldots, \widehat{\gamma_n}$ is an optimal solution to (3.3), we have either $\sum_{i \in A_j} \widehat{\gamma_i} = \pi_j$ or $\widehat{C_j} = 0$ (by complementary slackness). Thus, for any task T_j that is actually executed (i.e., $\widehat{C_j} > 0$), its total payoff $\pi_j \widehat{C_j}$ is shared only by players in A_j.

We now show that the outcome $(CS, \mathbf{x})$ is in the optimistic core. We can assume without loss of generality that CS allocates non-zero weight to the first k tasks $T_1, \ldots, T_k$ and no weight to the rest ($k \leq m$). Consider a deviation from $(CS, \mathbf{x})$ by a set S. This deviation can be described by a list of tasks that S abandons completely, and the amount of weight that players in S withdraw from all other tasks. Assume without loss of generality that the tasks that S abandons completely are $T_{\ell+1}, \ldots, T_k$ (this list includes all tasks T_j with $A_j \subseteq S$), and for each $j = 1, \ldots, \ell$ each member of $A_j \cap S$ withdraws z_j units of weight from T_j. Observe that "non-uniform" deviations are no better than "uniform" ones, i.e. if one player withdraws more weight from a coalition than the rest in an optimal deviation, then the rest may as well withdraw the same weight.

When players in S deviate, they lose their payoff from $T_{\ell+1}, \ldots, T_k$, and their payoff from $T_1, \ldots, T_\ell$ is reduced by $\sum_{j=1}^{\ell} z_j \pi_j$.

For each $i \in S$, set $\nu_i = \sum_{\ell < j \leq k, i \in A_j} \widehat{C_j}$, $Z_i = \sum_{j \leq \ell, i \in A_j} z_j$: ν_i is the total amount of weight that i withdraws from tasks $T_{\ell+1}, \ldots, T_k$ while Z_i is the total amount that i withdraws from $T_1, \ldots, T_\ell$. The profit that S obtains from optimally using the withdrawn resources is given by the following linear program:

$$\text{max:} \quad \sum_{A_j \subseteq S} C_j \pi_j \tag{3.4}$$
$$\text{s.t.} \quad \sum_{j: i \in A_j, A_j \subseteq S} C_j \leq \nu_i + Z_i \quad \forall i \in S$$

The dual of LP (3.4) is

$$\text{min:} \quad \sum_{i \in S} \gamma_i (\nu_i + Z_i) \tag{3.5}$$
$$\text{s.t.} \quad \sum_{i \in A_j} \gamma_i \geq \pi_j \quad \forall A_j \subseteq S$$

Let α be the value of (3.4) (and hence also of (3.5)). Note that the total profit that S gets by deviating equals $\alpha - \sum_{j=1}^{\ell} z_j \pi_j$, where $\sum_{j=1}^{\ell} z_j \pi_j$ is the total marginal loss incurred by S partially deviating from $T_1,\ldots,T_\ell$. Any optimal solution to (3.3) is a feasible solution to (3.5) (when looking at the restriction of the solution to those members of i); the constraints in (3.3) are more restricted than those in (3.5), since $\nu_i + Z_i \le W_i$ for all $i \in S$. Hence, given the optimal solution of the Dual (3.3) restricted to S, $(\widehat{\gamma}_i)_{i \in S}$, we have that

$$\alpha \le \sum_{i \in S} \widehat{\gamma}_i (\nu_i + Z_i).$$

Now, $\sum_{i \in S} \widehat{\gamma}^i \nu^i$ is exactly the payoff that S was getting from $T_{\ell+1},\ldots,T_k$ under $(CS, \mathbf{x})$. Further,

$$\begin{aligned}
\sum_{i \in S} \widehat{\gamma}_i Z_i &= \sum_{j=1}^{\ell} \sum_{i \in S \cap A_j} \widehat{\gamma}_i z_j \\
&= \sum_{j=1}^{\ell} z_j \left(\sum_{i \in S \cap A_j} \widehat{\gamma}_i \right) \le \sum_{j=1}^{\ell} z_j \left(\sum_{i \in A_j} \widehat{\gamma}_i \right) = \sum_{j=1}^{\ell} z_j \pi_j
\end{aligned}$$

where the last equality holds, since $\widehat{C}_j > 0$ for all $j = 1,\ldots,\ell$, and therefore, by complementary slackness, we have that $\sum_{i \in A_j} \widehat{\gamma}_i = \pi_j$. As previously mentioned, the latter expression is the marginal loss that S pays for withdrawing resources from $T_1,\ldots,T_\ell$. Thus, the total payoff that S gets from deviating is at most $\sum_{i \in S} \widehat{\gamma}_i \nu_i$; but:

$$\begin{aligned}
\sum_{i \in S} \widehat{\gamma}_i \nu_i &= \sum_{i \in S} \widehat{\gamma}_i \sum_{\ell < j \le k, i \in A_j} \widehat{c}_j \\
&\le \sum_{i \in S} \widehat{\gamma}_i \sum_{j : i \in A_j} \widehat{C}_j \\
&= \sum_{i \in S} p_i(CS, \mathbf{x}) = p_S(CS, \mathbf{x})
\end{aligned}$$

To conclude, the total payoff S receives from deviating under the optimistic arbitration function does not exceed its payoff in $(CS, \mathbf{x})$. As this holds for any deviation and any S, $(CS, \mathbf{x})$ is in the optimistic core. □

Finding an optimal solution for a linear program and its dual can be done in polynomial time; therefore, we obtain the following immediate corollary.

Corollary 3.23. *Given an LBG $\mathcal{G} = \langle N, \mathbf{W}, \mathcal{T} \rangle$,* OptVal, ArbVal, *and* Is-Stable *can be decided in time polynomial in n and $|\mathcal{T}|$ under the optimistic arbitration function.*

The running time guarantees provided in this section depend on both n and the number of tasks in the LBG. However, for some LBGs, the size of $\mathcal{T}$ is huge. For example, if one wishes to describe a network flow problem as an LBG, one would have a task for each directed path from the source to the target. In other words, the number of tasks could be exponential in the number of edges. In the specific instance of multicommodity flow games, this issue could be resolved via standard succinct representations of network flow games, as shown by Markakis and Saberi [2005]; in other cases, there is no such guarantee.

Since we know that the optimistic core is not empty for LBGs, it is in particular not empty for the conservative, sensitive and refined arbitration functions. In fact, if we assume that arbitration functions satisfy the accountability axiom, then Theorem 3.22 implies that the $\mathcal{A}$-core of an LBG is not empty for any arbitration function $\mathcal{A}$.

We also observe that the problem CHECKCORE is not resolved by our methods for LBGs. That is, given an outcome $(CS, \mathbf{x})$ for an LBG, it is not clear if deciding whether $(CS, \mathbf{x})$ is $\mathcal{A}$-stable can be done in polynomial time. We conjecture that this problem is computationally intractable, and leave this for future work.

As a final note, we observe that linear bottleneck games are 1-homogeneous: refactoring the resources of all players by a factor of α results in a change of α to the revenue generated. This, in particular, means that Corollary 2.26 applies to LBGs. When applied to LBGs, Corollary 2.26 is much weaker than Theorem 3.22: first, it only applies to the refined arbitration function, which is less generous than the optimistic arbitration function; second, it does not provide us with a poly-time algorithm for finding outcomes in the refined core, but simply uses the characterization of games with a non-empty refined core to show refined core non-emptiness.

Chapter 4

Alternative Solution Concepts in OCF Games

> "There is nobody in this country who got rich on their own. Nobody. You built a factory out there - good for you. But I want to be clear. You moved your goods to market on roads the rest of us paid for. You hired workers the rest of us paid to educate. You were safe in your factory because of police forces and fire forces that the rest of us paid for. You didn't have to worry that marauding bands would come and seize everything at your factory... Now look. You built a factory and it turned into something terrific or a great idea - God bless! Keep a hunk of it. But part of the underlying social contract is you take a hunk of that and pay forward for the next kid who comes along."
>
> —Elizabeth Warren

Chapters 2 and 3 have mostly been involved with one solution concept for OCF games, the arbitrated core. In this chapter we examine alternatives to the core.

It may seem unclear why would one need to define alternative solution concepts for OCF games, given the appeal and flexibilty of the arbitrated core. The drawbacks of the arbitrated core are similar to those of the classic core. First of all, it may be empty, i.e. for a given arbitration function, it is possible that there exist no outcomes that are resistant to deviations. Second of all, payoff divisions in the core may not be particularly fair; for example, they may offer different rewards to players who have contributed equally to profit generation. Finally, the core does not offer a unique solution, i.e., even if one can compute a core allocation in polynomial time, it is not clear which core allocation to choose.

Classic cooperative game theory handles these objections via alternative solution concepts. Some of the most popular ones are the Shapley value [Shapley, 1953], the nucleolus [Schmeidler, 1969], and the bargaining set [Davis and Maschler, 1963].

In this chapter, we propose three alternative solution concepts for OCF games. These three solution concepts are generalizations of the nucleolus, bargaining set and Shapley value for games with overlapping coalitions. We provide a formal definition and analysis of these solution concepts for OCF games. For the arbitrated nucleolus (Section 4.1), we show that it maintains many of the properties of its non-OCF counterpart. First, unlike the arbitrated core of OCF games, the arbitrated nucleolus is never empty; moreover, if the arbitrated core of a game is not empty, the arbitrated nucleolus is contained in the core (assuming that they use the same arbitration function), and is always contained in the arbitrated bargaining set. For the arbitrated bargaining set (Section 4.2), we show that the arbitrated core is contained in the arbitrated bargaining set, and that it is never empty (namely, it contains the arbitrated nucleolus). Finally, in Section 4.3, we define two possible extensions of the Shapley value for OCF games, both arising from two possible axiomatizations of a value for OCF games.

4.1 The Arbitrated Nucleolus

In this section, we extend the notion of the nucleolus [Schmeidler, 1969] to OCF games, and show that the resulting solution concept exhibits many of the desirable properties of its non-OCF counterpart. In a non-OCF cooperative game, the *nucleolus* of a cooperative game is an imputation that minimizes the unhappiness amongst players. Formally, given a non-OCF cooperative game $\mathcal{G} = \langle N, u \rangle$ and a vector $\mathbf{p} \in \mathbb{R}_+^n$ such that $p(N) = u(N)$, let $e(\mathbf{p}, S) = u(S) - p(S)$; $e(\mathbf{p}, S)$ is called the *excess* of S under $\mathbf{p}$. $e(\mathbf{p}, S)$ is a measure of S's unhappiness with $\mathbf{p}$; the greater the value of $e(\mathbf{p}, S)$ the more the set S is incentivized to deviate from $\mathbf{p}$. Let $\theta(\mathbf{p})$ be the vector $(e(\mathbf{p}, S_1), \ldots, e(\mathbf{p}, S_{2^n}))$, where $S_1, \ldots, S_{2^n}$ are all possible subsets of N, and are ordered such that $e(\mathbf{p}, S_1) \geq \cdots \geq e(\mathbf{p}, S_{2^n})$. The subsets of N are ordered with the unhappiest subset first, the second unhappiest second etc., with ties broken arbitrarily. The nucleolus is the set of imputations for which the vector $\theta(\mathbf{p})$ is minimal under the lexicographic ordering. It is shown in [Schmeidler, 1969] that

1. A nucleolus outcome is guaranteed to exist.

2. If the core is not empty, then the nucleolus is a member of the core.

3. The nucleolus consists of a single point.

When extending the definition of the nucleolus to OCF games, two factors come into play. First, one would like a definition that coincides with the non-OCF definition as much as possible; in particular, a definition of the nucleolus for OCF games should coincide with the nucleolus of its discrete superadditive cover in the case of the conservative arbitration function. This should be analogous to Theorem 2.11. In this section, we show a similar result for the arbitrated nucleolus.

Second, the excess of a set under a non-OCF cooperative game measures the difference between a set's bargaining power and payoff; i.e., $u(S)$ is the most a set S can stand to gain by deviating in the non-OCF setting. However, in an OCF game $\mathcal{G}$, the most a set stands to gain by deviating from a given outcome

Arbitration, Fairness and Stability

$(CS, \mathbf{x}) \in \mathcal{F}(\mathcal{G})$ is $\mathcal{A}^*(CS, \mathbf{x}, S)$. These two considerations lead us to the following definition of excess in OCF games.

Definition 4.1. Given an outcome $(CS, \mathbf{x}) \in \mathcal{F}(\mathcal{G})$, the *excess of a set* $S \subseteq N$ is defined as $e(CS, \mathbf{x}, S) = \mathcal{A}^*(CS, \mathbf{x}, S) - p_S(CS, \mathbf{x})$.

Much like in non-OCF games, the excess is a measure of a subset's "unhappiness" with a given outcome: the higher the excess, the unhappier the subset. Note also that $(CS, \mathbf{x})$ is in $Core(\mathcal{G}, \mathcal{A})$ if and only if $e(CS, \mathbf{x}, S) \leq 0$ for all $S \subseteq N$. Unlike classic cooperative games, however, the value of $\mathcal{A}^*$ may depend on the coalition structure that is chosen; in other words, it is possible that a coalition will have different excess under different coalition structures, a phenomenon that does not occur in classic cooperative games. Following the definitions given in [Schmeidler, 1969], given an outcome $(CS, \mathbf{x})$, we define its *excess vector* as

$$\theta(CS, \mathbf{x}) = (e(CS, \mathbf{x}, S_1), e(CS, \mathbf{x}, S_2), ..., e(CS, \mathbf{x}, S_{2^n})),$$

where $e(CS, \mathbf{x}, S_1) \geq ... \geq e(CS, \mathbf{x}, S_{2^n})$. We write $(CS, \mathbf{x}) \preceq_L (CS', \mathbf{y})$ if $\theta(CS, \mathbf{x})$ is lexicographically smaller than $\theta(CS', \mathbf{y})$. This analogous definition gives rise to an analogous definition of an arbitrated nucleolus.

Given an OCF game $\mathcal{G} = \langle N, v \rangle$ and an arbitration function $\mathcal{A}$, *the arbitrated nucleolus* of $\mathcal{G}$, denoted $Nuc(\mathcal{G}, \mathcal{A})$, is the set of all feasible outcomes that are minimal with respect to $\preceq_L$. Observe that just like in the non-overlapping case, if $Core(\mathcal{G}, \mathcal{A}) \neq \emptyset$, then $Nuc(\mathcal{G}, \mathcal{A}) \subseteq Core(\mathcal{G}, \mathcal{A})$.

4.1.1 Non-Emptiness of the Nucleolus

Unlike the arbitrated core, the nucleolus is never empty as long as $\mathcal{A}^*(CS, \mathbf{x}, S)$ is continuous over $\mathcal{F}(\mathcal{G})$, when S is fixed. Note that this property holds for all arbitration functions defined in Section 2.1.

Theorem 4.2. *If $\mathcal{A}^*$ is continuous over $\mathcal{F}(\mathcal{G})$, then $Nuc(\mathcal{G}, \mathcal{A}) \neq \emptyset$*

Proof. First, we would like to note that the excess vector is comprised of functions that achieve a minimum over $\mathcal{F}(\mathcal{G})$. Indeed, observe that for any outcome $(CS, \mathbf{x})$ and any $k = 1, ..., 2^n$ we have

$$\theta_k(CS, \mathbf{x}) = \max_{S_1, ..., S_k \subseteq N} \{\min\{e(CS, \mathbf{x}, S_1), ..., e(CS, \mathbf{x}, S_k)\}\},$$

where all $S_1, ..., S_k$ are different subsets of N. Since $\mathcal{A}^*$ is continuous, $e(CS, \mathbf{x}, S)$ has some minimal point in $\mathcal{F}(\mathcal{G})$. Thus, θ_k is obtained using a finite number of min and max operations, and therefore is continuous as well.

Set $X_1 = \arg\min_{(CS', \mathbf{y}) \in \mathcal{F}(\mathcal{G})}\{\theta_1(CS', \mathbf{y})\}$, and for every $k = 2, ..., 2^n$, let

$$X_k = \arg\min_{(CS', \mathbf{y}) \in X_{k-1}} \{\theta_k(CS', \mathbf{y})\}.$$

$X_{2^n} \subseteq Nuc(\mathcal{G}, \mathcal{A})$, since if $(CS, \mathbf{x}) \in X_k$ then $\theta_k(CS, \mathbf{x}) \leq \theta_k(CS', \mathbf{y})$ for every $k = 1, ..., 2^n$. Thus, it remains to show that X_{2^n} is non-empty. Now, the set $\mathcal{F}(\mathcal{G})$ is compact and non-empty. From elementary calculus, we know that if

$C \subseteq \mathbb{R}^m$ is a non-empty compact set, and $f : C \to \mathbb{R}$ is a continuous function, then the set $X = \{x \in C \mid f(x) = \min_{y \in C}\{f(y)\}\}$ is a non-empty compact set. Hence, X_1 is compact and non-empty, and inductively so is X_{2^n}. Consequently, $Nuc(\mathcal{G}, \mathcal{A}) \neq \emptyset$. □

Continuity of $\mathcal{A}^*$ is not strictly necessary in the proof of Theorem 4.2; we only require that $\mathcal{A}^*$ has a minimum over a compact subset of $\mathcal{F}(\mathcal{G})$ when S is fixed.

4.1.2 Properties of the Nucleolus

As we previously mention, the nucleolus in the non-OCF setting exhibits some attractive properties; for example, it is a single point. In the OCF setting, however, the nucleolus may have a richer structure.

Example 4.3. Consider the following TTG: $N = \{1, 2\}$. Both players have weight of 1 and there is one task t with $w(t) = 2, p(t) = 20$. Assume that G is arbitrated by the refined arbitrator. The r-core of the game is not empty and the only coalition structure that is in the r-core is $CS = \binom{1}{1}$. Let us consider a payoff distribution where player 1 gets $10 - \varepsilon$ and player 2 gets $10 + \varepsilon$ where $0 < \varepsilon < 10$. The maximum value that can be provided to player 1 under the refined arbitrator is if he offers $\binom{0}{0}$ as his objection; any other deviation will leave him with nothing. Indeed, $\mathcal{A}^*(CS, \mathbf{x}, \{1\}) = p_1(CS, \mathbf{x}) = 10 - \varepsilon$, thus his excess is 0. One can verify that all nucleolus outcomes have excess of 0 for all sets in this game. However, if the same game is arbitrated by the conservative or sensitive arbitrator, then the excess of player 1 is $0 - (10 - \varepsilon) = \varepsilon - 10$, which will make him sensitive to the fact that he is being cheated.

Example 4.3 demonstrates that outcomes in $Nuc(\mathcal{G}, \mathcal{A})$ need not be unique, nor distribute payoff among players in the same manner. However, it turns out that if $\mathcal{A}^*$ is convex as a function of $\mathbf{x}$ when S and CS are fixed, then for any two outcomes in the nucleolus that have the same coalition structure, each subset of players has the same excess under both of these outcomes. First, we prove the following technical lemma.

Lemma 4.4. *If* $(CS, \mathbf{x}), (CS, \mathbf{y}) \in Nuc(\mathcal{G}, \mathcal{A})$ *and* $\mathbf{z} = \frac{\mathbf{x}+\mathbf{y}}{2}$, *then* $(CS, \mathbf{z}) \in Nuc(\mathcal{G}, \mathcal{A})$.

Proof. Suppose that $(CS, \mathbf{x}), (CS, \mathbf{y})$ are both in $Nuc(\mathcal{G}, \mathcal{A})$. Both outcomes must have the same excess vector, i.e. $\theta(CS, \mathbf{x}) = \theta(CS, \mathbf{y})$. Set $\mathbf{z} = \frac{\mathbf{x}+\mathbf{y}}{2}$. Since $I(CS)$ is convex, $\mathbf{z}$ is also an imputation under CS, i.e. $\mathbf{z} \in I(CS)$. Indeed, consider the payoff to all players from some coalition $\mathbf{c}_j \in CS$:

$$\sum_{i=1}^{n} z_{ji} = \frac{1}{2}\sum_{i=1}^{n} x_{ji} + \frac{1}{2}\sum_{i=1}^{n} y_{ji} = v(\mathbf{c}_j).$$

Moreover, if $i \notin supp(\mathbf{c}_j)$ then $x_{ji} = y_{ji} = z_{ji} = 0$. Finally, if both $\mathbf{x}$ and $\mathbf{y}$ are individually rational, then $\mathbf{z}$ must be individually rational as well.

Consider $\theta(CS, \mathbf{z})$; let us write
$$\theta(CS, \mathbf{x}) = (e(CS, \mathbf{x}, J_1), ..., e(CS, \mathbf{x}, J_{2^n})),$$
$$\theta(CS, \mathbf{y}) = (e(CS, \mathbf{y}, K_1), ..., e(CS, \mathbf{y}, K_{2^n})),$$
$$\theta(CS, \mathbf{z}) = (e(CS, \mathbf{z}, L_1), ..., e(CS, \mathbf{z}, L_{2^n})).$$

Since $\mathcal{A}^*$ is convex, we have for all $S \subseteq N$
$$\mathcal{A}^*(CS, \mathbf{z}, S) \leq \frac{\mathcal{A}^*(CS, \mathbf{x}, S) + \mathcal{A}^*(CS, \mathbf{y}, S)}{2}.$$

Moreover,
$$\begin{aligned} p_S(CS, \mathbf{z}) &= \sum_{i \in S} p_i(CS, \mathbf{z}) \\ &= \sum_{i \in S} \sum_{j=1}^{m} z_{ji} \\ &= \frac{1}{2} \left(\sum_{i \in S} \sum_{j=1}^{m} x_{ji} + \sum_{i \in S} \sum_{j=1}^{m} y_{ji} \right) \\ &= \frac{1}{2} (p_S(CS, \mathbf{x}) + p_S(CS, \mathbf{y})) \end{aligned}$$

we conclude that
$$e(CS, \mathbf{z}, J) \leq \frac{1}{2} e(CS, \mathbf{x}, J) + \frac{1}{2} e(CS, \mathbf{y}, J). \tag{4.1}$$

Let us denote the l-th coordinate of $\theta(CS, \mathbf{x})$ by V_l, i.e. $e(CS, \mathbf{x}, J_l) = e(CS, \mathbf{y}, K_l) = V_l$. Using Equation 4.1 and setting $l = 1$ we get
$$e(CS, \mathbf{z}, L_1) \leq \frac{e(CS, \mathbf{x}, L_1) + e(CS, \mathbf{y}, L_1)}{2} \leq V_1.$$

If at any point the inequality is strict, $e(CS, \mathbf{z}, L_1) < V_1$ and $\theta(CS, \mathbf{z})$ is strictly smaller lexicographically than $\theta(CS, \mathbf{x})$, a contradiction. We similarly conclude that $e(CS, \mathbf{z}, L_k) = V_k$ for all $k = 1, \ldots, 2^n$. Therefore $\theta(CS, \mathbf{z}) = \theta(CS, \mathbf{x}) = \theta(CS, \mathbf{y})$, which means that $(CS, \mathbf{z})$ is in $Nuc(\mathcal{G}, \mathcal{A})$. □

Lemma 4.4 simply states that fixing a coalition structure CS, the set of nucleolus outcomes whose coalition structure is CS is convex.

Theorem 4.5. *Let $\mathcal{G} = \langle N, v \rangle$ be a game arbitrated by some convex arbitrator $\mathcal{A}$. If $(CS, \mathbf{x})$ and $(CS, \mathbf{y})$ are in $Nuc(\mathcal{G}, \mathcal{A})$ then for any $S \subseteq N$ we have $e(CS, \mathbf{x}, S) = e(CS, \mathbf{y}, S)$.*

Proof. The proof scheme is somewhat similar in flavor to the proof that the nucleolus for non-OCF games is unique (see [Maschler et al., 2013] for details). Let $(CS, \mathbf{x}), (CS, \mathbf{y})$ be in $Nuc(\mathcal{G}, \mathcal{A})$. Set $\mathbf{z} = \frac{\mathbf{x}+\mathbf{y}}{2}$. Using the same notation as in Lemma 4.4, we know that $e(CS, \mathbf{x}, J_1)$ equals $e(CS, \mathbf{y}, K_1)$ and $e(CS, \mathbf{z}, L_1)$, so
$$e(CS, \mathbf{x}, J_1) + e(CS, \mathbf{y}, K_1) = 2e(CS, \mathbf{z}, L_1). \tag{4.2}$$

As shown in Lemma 4.4,
$$2e(CS, \mathbf{z}, L_1) \le e(CS, \mathbf{x}, L_1) + e(CS, \mathbf{y}, L_1). \tag{4.3}$$

By definition of J_1, $e(CS, \mathbf{x}, L_1) \le e(CS, \mathbf{x}, J_1)$ and similarly, $e(CS, \mathbf{y}, L_1) \le e(CS, \mathbf{y}, K_1)$. Thus, on the one hand, by Equation (4.2) we have that
$$2e(CS, \mathbf{z}, L_1) \ge e(CS, \mathbf{x}, L_1) + e(CS, \mathbf{y}, L_1),$$

Combining the above with Equation (4.3), we get that
$$e(CS, \mathbf{x}, L_1) + e(CS, \mathbf{y}, L_1) = e(CS, \mathbf{x}, J_1) + e(CS, \mathbf{y}, K_1).$$

If $e(CS, \mathbf{x}, L_1) < e(CS, \mathbf{x}, J_1)$, then $e(CS, \mathbf{y}, L_1)$ must be strictly greater than $e(CS, \mathbf{y}, K_1)$, which is impossible, thus
$$e(CS, \mathbf{y}, K_1) = e(CS, \mathbf{y}, L_1) = e(CS, \mathbf{x}, J_1) = e(CS, \mathbf{x}, L_1).$$

This means that we can swap L_1 with J_1 in the excess ordering of $(CS, \mathbf{x})$ without changing the excess vector $\theta(CS, \mathbf{x})$. This can be done inductively for any L_k. We conclude that if $(CS, \mathbf{x}), (CS, \mathbf{y})$ are in $Nuc(\mathcal{G}, \mathcal{A})$ then all sets have the same excess in both outcomes. □

We can obtain a slightly stronger version of Theorem 4.5 for the conservative arbitration function (or for that matter, any arbitration function where $\mathcal{A}^*$ does not depend on CS).

Corollary 4.6. *If $\mathcal{G} = \langle N, v \rangle$ is arbitrated by the conservative arbitration function $\mathcal{A}_c$, then for any $(CS, \mathbf{x}), (CS, \mathbf{y}) \in Nuc(\mathcal{G}, \mathcal{A}_c)$ and any $i \in N$, $p_i(CS, \mathbf{x}) = p_i(CS, \mathbf{y})$.*

Proof. Consider two outcomes $(CS, \mathbf{x}), (CS, \mathbf{y}) \in Nuc(\mathcal{G}, \mathcal{A}_c)$ and a player i. First, observe that $\mathcal{A}_c^*(CS, \mathbf{x}, S) = v^*(\mathbf{e}^S)$ for all $S \subseteq N$. In particular, for any $i \in N$, and any outcome $(CS, \mathbf{x}) \in Nuc(\mathcal{G}, \mathcal{A}_c)$ we have $e(CS, \mathbf{x}, \{i\}) = v^*(\mathbf{e}^{\{i\}}) - p_i(CS, \mathbf{x})$. Let $\mathbf{x}, \mathbf{y} \in I(CS)$ be imputations such that $(CS, \mathbf{x})$ and $(CS, \mathbf{y})$ are in $Nuc(\mathcal{G}, \mathcal{A}_c)$; according to Theorem 4.5, we have $e(CS, \mathbf{x}, \{i\}) = e(CS, \mathbf{y}, \{i\})$ which implies that $p_i(CS, \mathbf{x}) = p_i(CS, \mathbf{y})$. □

We remark that one can also show that the refined arbitrator is convex; however, as illustrated by Example 4.3, the conclusion of Corollary 4.6 does not hold for the refined arbitrator, since the value of $\mathcal{A}_r^*(CS, \mathbf{z}, J)$ may depend on the vector $\mathbf{z}$. Also, as shown in [Aumann and Drèze, 1974], two coalition structures may induce different nucleolus outcomes, even in the non-OCF setting. Thus, while Corollary 4.6 shows that player payoffs do not depend on the imputation chosen once the coalition structure is fixed, they do depend on the coalition structure chosen.

4.2 The Arbitrated Bargaining Set

The *bargaining set* of a non-OCF cooperative game is the set of all imputations to which no player can justifiably object. The idea is that an agent will only agree to deviate if he knows that his deviation is resistant to outside influence. Formally, given a non-OCF cooperative game $\mathcal{G} = \langle N, u \rangle$ and an imputation **p**, an *objection* of $i \in N$ against $j \in N$ under **p**, is a subset $S \subseteq N \setminus \{i,j\}$ and a payoff division **q** of $u(S \cup \{i\})$ such that $q_k > p_k$ for all $k \in S$, and $q_i > p_i$. That is, player i can take a subset of players in N that does not include j, form a coalition with them, and ensure that they receive a strictly better payoff than what they receive under **p**. A *counter-objection* of j to an objection of i to j is a set $T \subseteq N \setminus \{i,j\}$, and a payoff division **r** of $u(T \cup \{j\})$. The payoff division **r** is such that

1. for all $k \in T \cap S$, $r_k \geq q_k$.

2. for all $k \in T \setminus S$, $r_k \geq p_k$.

3. $r^j \geq p^j$.

Thus, j can counter-object to i if he can block i's original objection by offering a (weakly) better deal to all players who agreed to deviate with i, while ensuring that any new members (i.e. $T \setminus S$), and himself, are not worse off. An objection of i to j is called *justified* if j cannot counter-object to i.

The *bargaining set* [Davis and Maschler, 1963] of a non-OCF cooperative game is the set of all imputations to which no justified objections exist. We note that no player can object to a core imputation; thus if the core of a non-OCF game is not empty, then it is contained in the bargaining set. Moreover, it is known [Davis and Maschler, 1963] that the nucleolus is a point in the bargaining set of a game, which means in particular that the bargaining set is never empty.

Having defined the nucleolus of an arbitrated OCF game, it is only natural to define the bargaining set of $\mathcal{G}$. This is the set of all outcomes which no player can justifiably object to in our setting. We use a definition of the bargaining set similar in essence to the one defined by Davis and Maschler [1963]. Suppose we are given an outcome in $\mathcal{F}(\mathcal{G})$. Different subsets will have different opinions about the way payoff was distributed and coalitions were formed. One can picture this as a process where players attempt to threaten and object to each other, until some fair outcome is decided upon. We begin by formally defining an objection of a player in the OCF setting.

Definition 4.7. An objection of a player i to a player j under an outcome $(CS, \mathbf{x})$ is an $\mathcal{A}$-profitable deviation of a subset $S_i \subseteq N$ where $i \in S_i$ but $j \notin S_i$. We say that i objects to j with $(CS', \mathbf{y})$ and $\mathbf{r} = (\mathbf{r}(\mathbf{c}))_{\mathbf{c} \in CS \setminus CS|_{S_i}}$ if $(CS', \mathbf{y})$ is the outcome under which S_i $\mathcal{A}$-profitably deviates, and **r** describes the way arbitration payoffs from coalitions that are not in $CS|_{S_j}$ are divided among the members of S_i.

If $(CS, \mathbf{x}) \in Core(\mathcal{G}, \mathcal{A})$, then no player i can object to any player j under $(CS, \mathbf{x})$. This is true since no subset $S \subseteq N$ can $\mathcal{A}$-profitably deviate from $(CS, \mathbf{x})$.

Definition 4.8. Suppose that i objected to j with S_i, using $(CS', \mathbf{y})$ and **r**. We say that j has a counter objection to i if there is some $S_j \subseteq N \setminus \{i\}$ and a deviation of S_j such that all members of $S_j \cap S_i$ receive weakly more than what they had received under $(CS', \mathbf{y})$ and $S_j \setminus S_i$ receive weakly more than what they had received under $(CS, \mathbf{x})$ under the deviation by S_j; finally, after S_j deviates, j must receive weakly more than what he does under $(CS, \mathbf{x})$.

The last point in Definition 4.8 is important to stress here. Player j need not actually deviate in order to successfully counter object to i. It suffices that j points out a set not containing i, S_j, such that all players in $S_j \cap S_i$ receive weakly more than what i offers, and $S_j \setminus S_i$ are weakly better off than under $(CS, \mathbf{x})$, and such that j himself is not hurt by the deviation.

If j cannot offer a counter objection, we say that i's objection is *justified*. Intuitively, if j can object to i, then j can offer a payoff scheme that will have her and some collaborators weakly better off, and will provide members of $S_i \cap S_j$ with a weakly better offer than the one they received from i. Having a well-defined notion of objections and counter-objections for OCF games, we are ready to define the bargaining set of an OCF game.

Definition 4.9. The bargaining set of a game $\mathcal{G} = \langle N, v \rangle$ given $\mathcal{A}$ is the set of all feasible outcomes that no player i can justifiably object to some other player j under. we denote the bargaining set of $\mathcal{G}$ as $Barg(\mathcal{G}, \mathcal{A})$.

Since no player can object to any other player under any $(CS, \mathbf{x}) \in Core(\mathcal{G}, \mathcal{A})$, in particular no justified objections exist to such outcomes. Therefore, $Core(\mathcal{G}, \mathcal{A}) \subseteq Barg(\mathcal{G}, \mathcal{A})$.

4.2.1 Non-Emptiness of the Bargaining Set

We now show that the bargaining set is never empty. We do this by showing that the nucleolus of an OCF game is always contained in the bargaining set. Our proof is similar in vein to the one found in [Maschler et al., 2013] for non-OCF games. We begin by proving some auxiliary lemmas.

Lemma 4.10. *Suppose $i \in N$ has a justified objection to $j \in N$ under $(CS, \mathbf{x})$ with some subset $S_i \subseteq N$ containing i but not j. Then we have:*

$$e(CS, \mathbf{x}, S_j) < e(CS, \mathbf{x}, S_i)$$

for any $S_j \subseteq N$ containing j but not i.

Proof. Assume that for some $S_j \subseteq N$ containing j but not i we have $e(CS, \mathbf{x}, S_j) \geq e(CS, \mathbf{x}, S_i)$. Suppose that i objected to j with a subset S_i, an outcome $(CS', \mathbf{y})$ and a division of arbitration payoffs $\mathbf{r}$. Let us write q_ℓ to be the total payoff to $\ell \in S_i$ under the deviation.

Observe the total payoff to $S_i \cup S_j$ under $(CS, \mathbf{x})$, $p_{S_i \cup S_j}(CS, \mathbf{x})$. We know that every $\ell \in S_i$ strictly gains from the deviation offered by i, so $p_{S_i \cup S_j}(CS, \mathbf{x}) < q(S_i \setminus S_j) + p_{S_j}(CS, \mathbf{x})$. Since $p_{S_j \cup S_i}(CS, \mathbf{x}) = p_{S_i}(CS, \mathbf{x}) + p_{S_j \setminus S_i}(CS, \mathbf{x})$ we get:

$$p_{S_j}(CS, \mathbf{x}) - p_{S_i}(CS, \mathbf{x}) > p_{S_j \setminus S_i}(CS, \mathbf{x}) - q(S_i \setminus S_j)$$

Since $e(CS, \mathbf{x}, S_j) \geq e(CS, \mathbf{x}, S_i)$ we know that $\mathcal{A}^*(CS, \mathbf{x}, S_j) - p_{S_j}(CS, \mathbf{x}) \geq \mathcal{A}^*(CS, \mathbf{x}, S_i) - p_{S_i}(CS, \mathbf{x})$, or:

$$\mathcal{A}^*(CS, \mathbf{x}, S_j) - \mathcal{A}^*(CS, \mathbf{x}, S_i) \geq p_{S_j}(CS, \mathbf{x}) - p_{S_i}(CS, \mathbf{x})$$

Combining the two inequalities we get that

$$\mathcal{A}^*(CS, \mathbf{x}, S_j) - \mathcal{A}^*(CS, \mathbf{x}, S_i) > p_{S_j \setminus S_i}(CS, \mathbf{x}) - q(S_i \setminus S_j).$$

Since q is the payoff from an $\mathcal{A}$-profitable deviation from $(CS, \mathbf{x})$ we know that $q(S_i) \leq \mathcal{A}^*(CS, \mathbf{x}, S_i)$. So we have:

$$\mathcal{A}^*(CS, \mathbf{x}, S_i) \geq q(S_i) = q(S_i \cap S_j) + q(S_i \setminus S_j).$$

We get that:

$$\mathcal{A}^*(CS, \mathbf{x}, S_j) - \mathcal{A}^*(CS, \mathbf{x}, S_i) \leq \mathcal{A}^*(CS, \mathbf{x}, S_j) - q(S_i \cap S_j) - q(S_i \setminus S_j).$$

Combining the inequalities we get:

$$\mathcal{A}^*(CS, x, S_j) - q(S_i \cap S_j) - q(S_i \setminus S_j) > p_{S_j \setminus S_i}(CS, \mathbf{x}) - q(S_i \setminus S_j),$$

or:

$$\mathcal{A}^*(CS, \mathbf{x}, S_j) > q(S_i \cap S_j) + p_{S_j \setminus S_i}(CS, \mathbf{x}).$$

We now use a similar method to that used in Theorem 2.3 in order to show that there exists some non-empty subset $K \subseteq S_j$ such that every $\ell \in K \cap S_i$ receives strictly more than q_ℓ by deviating, and that every member of $K \setminus S_i$ receives weakly more than $p_\ell(CS, \mathbf{x})$ by deviating. Moreover, all members of $S_j \setminus K$ receive the same payoff as before. Let us choose some deviation CS'' of S_j from CS such that the total payoff to S_j from deviating is maximized. Let CS_d be the coalition structure that S_j forms using the resources it withdrew from $CS \setminus CS|_S$ via CS'', as well as the resources in $CS|_S$; i.e. $v(CS_d) = v^*(\mathbf{w}(CS|_{S_j}) + \mathbf{w}_{S_j}(CS''))$. Moreover, let $\alpha_\mathbf{c}$ be the payoff to S_j from the coalition $\mathbf{c} \in CS \setminus CS|_S$ given CS'', under $\mathcal{A}$. Given $\alpha_\mathbf{c}$, let $I_\mathbf{c}$ be the set of all possible payoff divisions given $\alpha_\mathbf{c}$; in other words, if $\alpha_\mathbf{c} \geq 0$ then

$$I_\mathbf{c} = \{\mathbf{y} \in \mathbb{R}_+^n \mid supp(\mathbf{y}) \subseteq S_j; \sum_{\ell \in S_j} y^\ell = \alpha_\mathbf{c}\},$$

and if $\alpha_\mathbf{c} < 0$ then

$$I_\mathbf{c} = \{\mathbf{y} \in \mathbb{R}^n \mid supp(\mathbf{y}) \subseteq S_j; \sum_{\ell \in S_j} y_\ell = \alpha_\mathbf{c}; y_\ell \leq 0 \text{ for all } \ell \in S_j\}.$$

We write $I = \prod_{\mathbf{c} \in CS \setminus CS|_{S_j}} I_\mathbf{c}$. Given some $\mathbf{y} \in I(CS_d)$ and some $\mathbf{r} \in I$, let us write $q_\ell(\mathbf{y}, \mathbf{r})$ to be the total payoff to $\ell \in S_j$ under $\mathbf{y}$ and $\mathbf{r}$. We again write

$$TL(\mathbf{y}, \mathbf{r}) = \sum_{\ell \in S_j : q_\ell(\mathbf{y},\mathbf{r}) > p_\ell(CS, \mathbf{x})} q_\ell(\mathbf{y}, \mathbf{r}) - p_\ell(CS, \mathbf{x}),$$

the total marginal loss caused under $\mathbf{y}$ and $\mathbf{r}$. Since TL is continuous and both Δ and $I(CS_d)$ are compact, there exist some $\bar{\mathbf{y}}$ and $\bar{\mathbf{r}}$ that minimize TL over $I(CS_d) \times \Delta$. We choose $\bar{\mathbf{y}}$ and $\bar{\mathbf{r}}$ that maximize the number of players for which $p_\ell(CS, \mathbf{x}) < q_\ell(\bar{\mathbf{y}}, \bar{\mathbf{r}})$, i.e. the number of players who strictly gain by deviating is maximized.

We construct a directed graph as follows. The players in S_j are nodes, and there is an edge from ℓ to k if one of the following holds:

1. There is some coalition $\mathbf{c} \in CS_d$ such that $y_\ell(\mathbf{c}) > 0$ and $\ell, k \in supp(\mathbf{c})$.

2. There is some coalition $\mathbf{c} \in CS \setminus CS|_S$ such that $\alpha_\mathbf{c} > 0$, $r_\ell(\mathbf{c}) > 0$ and $\ell, k \in supp(\mathbf{c})$.

3. There is some coalition $\mathbf{c} \in CS \setminus CS|_S$ such that $\alpha_\mathbf{c} < 0$, $r_k(\mathbf{c}) < 0$ and $\ell, k \in supp(\mathbf{c})$.

In other words, there is an edge from ℓ to k if ℓ can transfer some of his payoff to k without violating coalitional constraints (or, in the case of a cost incurred, that ℓ can assume some of the cost incurred by k). We color a vertex ℓ green if $p_\ell(CS, \mathbf{x}) < q_\ell(\bar{\mathbf{y}}, \bar{\mathbf{r}})$, red if $p_\ell(CS, \mathbf{x}) > q_\ell(\bar{\mathbf{y}}, \bar{\mathbf{r}})$, and white otherwise.

We note that under the assumption on $\bar{\mathbf{y}}$ and $\bar{\mathbf{r}}$, there is no edge from a green vertex to a non-green vertex. If there is an edge from a green vertex $g \in S_j$ to a red vertex r, then g can transfer a small amount of payoff to r while still remaining green, thus resulting in a payoff division with a strictly smaller value of TL, a contradiction to the minimality of $\bar{\mathbf{y}}$ and $\bar{\mathbf{r}}$. If there is an edge from a green vertex to a white vertex, then by transferring a small payoff to the white vertex, he will become green as well. This is a contradiction to the fact that the number of green vertices is maximal under $\bar{\mathbf{y}}$ and $\bar{\mathbf{r}}$.

We conclude that under $(\bar{\mathbf{y}}, \bar{\mathbf{r}})$, if a green player is paid by some coalition $\mathbf{c}$, then the support of $\mathbf{c}$ contains only green players. If a green and non-green player are in the same coalition, then the green player receives no payoff from that coalition. Let us write $S_g \subseteq S_j$ to be the set of all green players, then $S_g \neq \emptyset$ since the total payoff to S_j by deviating exceeds the total payoff it receives if it does not. Moreover, if we assume that only S_g deviates, while $S_j \setminus S_g$ sticks to the original plan, then all members of S_g will still receive at least the payoff they receive before: this is due to the deviation monotonicity property of arbitration functions. Therefore, S_g can deviate from $(CS, \mathbf{x})$, ensuring that every member $\ell \in S_g \cap S_i$ receives more than q_ℓ, and every member of $S_g \setminus S_i$ receives more than $p_\ell(CS, \mathbf{x})$.

If $j \in S_g$ we are done: j can object to i using S_g and we arrive at a contradiction. Otherwise, recall that any change in the actions of members outside of S_g will not affect G's payoff. Therefore, if members of $S_j \setminus S_g$ keep to their original strategy under CS, they will receive the same payoff. Thus j can counter-object to i using S_j, and we again reach a contradiction. □

Lemma 4.10 states that if i has a justified objection to j, then all subsets containing j but not i are happier than all subsets containing i but not j. We now show that if $\mathcal{A}$ satisfies the accountability property (i.e. $\mathcal{A} \leq \mathcal{A}_o$), then justified objections to nucleolus outcomes can only exist if some non-deviators interact in a meaningful way with deviators.

Lemma 4.11. *Consider some $(CS, \mathbf{x})$, where CS is an optimal coalition structure, and an $\mathcal{A}$-profitable deviation of S from $(CS, \mathbf{x})$; if $\mathcal{A}$ satisfies accountability, then there exists some $j \in N \setminus S$ and some $\mathbf{c} \in CS$ such that*

(a) $S \cap supp(\mathbf{c}) \neq \emptyset$.

(b) *Player j is in $supp(\mathbf{c})$.*

(c) *j receives some positive payoff from $\mathbf{c}$.*

Proof. If this is not the case, then the payoff to $N \setminus S$ is received solely from $CS|_{N \setminus S}$. Now, suppose that after deviating according to a deviation $(\mathbf{d}(\mathbf{c}))_{\mathbf{c} \in CS \setminus CS|_S}$, S forms the coalition structure CS' and divides profits from CS' according to $\mathbf{y} \in I(CS')$. Recall that $\mathcal{A} \leq \mathcal{A}_o$ (i.e. S receives at most the payoff it would have received under the optimistic arbitration function); thus, for any coalition $\mathbf{c} \in CS$ such that $supp(\mathbf{c}) \cap S \cap (N \setminus S) \neq \emptyset$, $\alpha_\mathbf{c} \leq \max\{0, v(\mathbf{c} - \mathbf{d}(\mathbf{c})) - \sum_{i \in N \setminus S} x_i(\mathbf{c})\}$. We also write $\mathbf{r}(\mathbf{c})$ to be the way that players from S divide the rewards (or fines) from the coalition $\mathbf{c}$, $\alpha_\mathbf{c}$.

Let us define the coalition structure $\bar{CS}$ to be the coalitions that players form post deviation; that is,

$$\bar{CS} = (\mathbf{c} - \mathbf{d}(\mathbf{c}))_{\mathbf{c} \in CS \setminus CS|_S} \cup CS'.$$

We define the imputation $\bar{\mathbf{x}}$ to be as follows: for every coalition $\mathbf{c}' \in CS'$ we set $\bar{\mathbf{x}}(\mathbf{c}') = \mathbf{y}(\mathbf{c}')$; for every coalition $\bar{\mathbf{c}} = \mathbf{c} - \mathbf{d}(\mathbf{c})$ where $\mathbf{c} \in CS \setminus CS|_S$, we let $\bar{\mathbf{x}}(\bar{\mathbf{c}})$ to be as follows: if $\alpha_\mathbf{c} < 0$ then $\bar{\mathbf{x}}(\bar{\mathbf{c}})$ can be any profit division of $v(\mathbf{c} - \mathbf{d}(\mathbf{c}))$ among the members of S. Otherwise, $\bar{\mathbf{x}}(\bar{\mathbf{c}})$ can be any division of profits among the members of S, so long as every $i \in S \cap supp(\bar{\mathbf{c}})$ receives at least $r_i(\mathbf{c})$, and every $i \in N \setminus S$ receives 0 (the payoff to all members of $N \setminus S$ from such a coalition is 0 under $(CS, \mathbf{x})$ by assumption). We can pay each player in $S \cap supp(\bar{\mathbf{c}})$ at least $r_i(\mathbf{c})$, since

$$\sum_{i \in S \cap supp(\bar{\mathbf{c}})} r_i(\mathbf{c}) = \alpha_\mathbf{c} \leq v(\mathbf{c} - \mathbf{d}(\mathbf{c})) - \sum_{i \in N \setminus S} x_i(\mathbf{c}) = v(\mathbf{c} - \mathbf{d}(\mathbf{c})) = v(\bar{\mathbf{c}}).$$

Now, under $(\bar{CS}, \bar{\mathbf{x}})$, S receives at least what it gets by deviating, so $p_S(\bar{CS}, \bar{\mathbf{x}}) > p_S(CS, \mathbf{x})$, and $p_{N \setminus S}(\bar{CS}, \bar{\mathbf{x}}) = p_{N \setminus S}(CS, \mathbf{x})$. Therefore, $v(\bar{CS}) > v(CS)$, a contradiction to CS being an optimal coalition structure. □

Lemma 4.12. *If i can object to j using some subset S_i, then any $i' \in S_i$ has an objection to j; if i can object to j using some subset S_i, then i can object to any j' in $N \setminus S_i$.*

Proof. Since the payoff to all of S_i's members under the deviation of S_i from $(CS, \mathbf{x})$ is strictly better off compared to $(CS, \mathbf{x})$, then any $i' \in S_i$ can object to j using S_i and offering the same deviation. Similarly, since any $j' \in N \setminus S_i$ is (by definition), not in S_i, then i's objection to j is also an objection to j'. □

Using the three lemmas above, we are now able to prove the following result.

Theorem 4.13. $Nuc(\mathcal{G}, \mathcal{A}) \subseteq Barg(\mathcal{G}, \mathcal{A})$.

Proof. If $Core(\mathcal{G}, \mathcal{A}) \neq \emptyset$ then we have $Nuc(\mathcal{G}, \mathcal{A}) \subseteq Core(\mathcal{G}, \mathcal{A}) \subseteq Barg(\mathcal{G}, \mathcal{A})$ and we are done. We thus assume that $Core(\mathcal{G}, \mathcal{A}) = \emptyset$. This means that for every outcome $(CS, \mathbf{x}) \in \mathcal{F}(\mathcal{G})$ there exists some $S \subseteq N$ such that S can $\mathcal{A}$-profitably deviate from $(CS, \mathbf{x})$. Let $(CS, \mathbf{x}) \in Nuc(\mathcal{G}, \mathcal{A})$. We must have then that $\theta_1(CS, \mathbf{x}) > 0$ since $Core(\mathcal{G}, \mathcal{A}) = \emptyset$.

Assume now that there is some $(CS, \mathbf{x}) \in Nuc(\mathcal{G}, \mathcal{A})$ that is not in the bargaining set of $\mathcal{G}$; so some player i can justifiably object to some player j with an outcome $(CS', \mathbf{y})$ and a subset S_i not containing j. Combining Lemma 4.11 and

Lemma 4.12, we can assume that i and j share a coalition where j gets a positive payoff. We denote the coalition that i and j shared where j's payoff was positive as **c**. Since S_i can profitably deviate from $(CS, \mathbf{x})$, we have that $e(CS, \mathbf{x}, S_i) > 0$.

According to Lemma 4.10, every subset $S \subseteq N \setminus \{i\}$ containing j must have lower excess than $e(CS, \mathbf{x}, S_i)$. Let I_M be the subset of $N \setminus \{j\}$ containing i with the highest excess. Denote $e(CS, \mathbf{x}, I_M) = a$. We order all those coalitions $J_1, ..., J_m$ containing j but not i according to their excess. Every J_k has a lower excess than a. We denote $e(CS, \mathbf{x}, J_1) = b$. Note that for any subset $S \subseteq N$ with a higher excess we must have that S contains either both i and j or neither. Since i and j belong to the same coalition **c** and $x_j(\mathbf{c}) > 0$, j can legally transfer payoff to i.

We define $\lambda = \min\{a - b, x_j(\mathbf{c})\}$; since $x_j(\mathbf{c}) > 0$, and $a > b$, λ is positive. We choose some ε such that $0 < \varepsilon < \frac{\lambda}{2}$.

By transferring an amount of ε from player j to player i (under the coalition **c**), we get that the excess has decreased for at least one subset with excess a, and for all those above it in the ordering it did not change as both i and j are there, or neither of them. Since $\varepsilon < a - b$ no subset containing j has increased its excess to be more than a. This results in an outcome that is strictly lower lexicographically than $(CS, \mathbf{x})$, a contradiction to $(CS, \mathbf{x})$ being in $Nuc(\mathcal{G}, \mathcal{A})$. □

As an immediate corollary of Theorem 4.13, we have that $Barg(\mathcal{G}, \mathcal{A}) \neq \emptyset$.

4.3 The Shapley Value in OCF Games

Introduced by Shapley [1953], the Shapley value is a central solution concept in classic cooperative game theory. Let us first recall the definition of the Shapley value for non-OCF games. Given a non-OCF cooperative game $\mathcal{G} = \langle N, u \rangle$, let $m_i(S)$ be the *marginal contribution* of a player $i \in N$ to the subset $S \subseteq N$; that is, $m_i(S) = u(S \cup \{i\}) - u(S)$. Let $\Pi(N)$ be the set of all bijective functions from N to itself i.e. $\Pi(N)$ is the set of permutations, or orderings, of N. Given a permutation $\sigma \in \Pi(N)$ and some $i \in N$, let $P_i(\sigma)$ be the set of the *predecessors* of i in σ; that is

$$P_i(\sigma) = \{j \in N \mid \sigma(j) < \sigma(i)\}.$$

We define the marginal contribution of i to σ as the marginal contribution of i to his predecessors; we write

$$m_i(\sigma) = m_i(P_i(\sigma)) = u(P_i(\sigma) \cup \{i\}) - u(P_i(\sigma)).$$

Suppose that we pick a permutation $\sigma \in \Pi(N)$ uniformly at random; let $X_i(\sigma)$ be the random variable which equals $m_i(\sigma)$; the *Shapley value* of player i is $\mathbb{E}[X_i(\sigma)]$, where $\mathbb{E}[X]$ is the expected value of a random variable X. Simply put, the Shapley value of player i is his expected marginal contribution to a randomly selected permutation. We write

$$sv_i(\mathcal{G}) = \mathbb{E}[X_i(\sigma)] = \frac{1}{n!} \sum_{\sigma \in \Pi(N)} m_i(\sigma).$$

We write sv_i to denote the Shapley value of player i in $\mathcal{G}$ when the underlying game is clear. The appeal of the Shapley value as a payoff division in cooperative games stems from the fact that it can be derived axiomatically; Shapley [1953] shows that the payoff division induced by giving each player a payoff of sv_i is the only payoff division satisfying several desirable properties. We are interested in families of values (i.e. solution concepts that consist of a single point) that satisfy certain axioms. Given a value Φ of a game $\mathcal{G}$, we refer to the i-th coordinate of Φ as $\Phi_i(\mathcal{G})$, or Φ_i when $\mathcal{G}$ is clear from the context. We begin by describing the following axioms.

Efficiency: Φ is *efficient* if $\sum_{i=1}^{n} \Phi_i = u(N)$.

Symmetry: Φ is *symmetric* if for any two symmetric players $i, j \in N$, $\Phi_i = \Phi_j$. Players i and j are symmetric if $m_i(S) = m_j(S)$ for all $S \subseteq N \setminus \{i, j\}$.

Dummy: a player $i \in N$ is called a *dummy* if $m_i(S) = 0$ for all $S \subseteq N$; Φ has the dummy property if $\Phi_i = 0$ for all dummy players.

Additivity: Given two games $\mathcal{G} = \langle N, u \rangle$ and $\bar{\mathcal{G}} = \langle N, \bar{u} \rangle$, let $\mathcal{G} + \bar{\mathcal{G}} = \langle N, \hat{u} \rangle$ be a game such that for all $S \subseteq N$, $\hat{u}(S) = u(S) + \bar{u}(S)$. We say that Φ is *additive* if for all $\mathcal{G}$ and $\bar{\mathcal{G}}$, $\Phi_i(\mathcal{G}) + \Phi_i(\bar{\mathcal{G}}) = \Phi_i(\mathcal{G} + \bar{\mathcal{G}})$.

Strong Monotonicity: given two games $\mathcal{G} = \langle N, u \rangle$ and $\bar{\mathcal{G}} = \langle N, \bar{u} \rangle$, suppose that for some $i \in N$ we have that for all $S \subseteq N$, $u(S \cup \{i\}) - u(S) \geq \bar{u}(S \cup \{i\}) - \bar{u}(S)$; that is, the marginal contribution of i to all S is greater under $\mathcal{G}$ than under $\bar{\mathcal{G}}$.

Shapley [1953] has shown the first axiomatization of the Shapley value.

Theorem 4.14. *[Shapley [1953]] $sv(\mathcal{G})$ is the only value satisfying the efficiency, symmetry, dummy and additivity axioms.*

An alternative characterization of the Shapley value has appeared in [Young, 1985]; it uses strong monotonicity rather than additivity.

Theorem 4.15. *[Young [1985]] $sv(\mathcal{G})$ is the only value satisfying the efficiency, symmetry and strong monotonicity axioms.*

We note that there are other axiomatizations of the Shapley value, as well as axiomatic approaches to the definition of the core, the nucleolus and the bargaining set. See Peleg and Sudhölter [2007] for a detailed review, as well as Winter [2002] for review of axiomitizations of the Shapley value.

We offer two possible extensions of the Shapley value to OCF games; one assumes a fixed coalition structure and is somewhat similar to the Shapley value for coalition structures defined in [Aumann and Drèze, 1974], while the other takes into account the ability of sets to maximize their profits using coalition structures and is similar to the classic notion defined in [Shapley, 1953]. We show that both values are unique with regard to specific sets of axioms, similar to those defined above.

Our first definition assumes that a coalition structure $CS \in \mathcal{CS}(N)$ is given. The motivation for this scenario is as follows: suppose that players have formed a coalition structure by some means; it may have been specified by some algorithm or a central authority, or agreed upon by the players themselves. Having formed the coalition structure CS, players are now interested in assessing the contribution of each player to the coalition structure. Note that this type of assessment does not measure the intrinsic value of a player to the team; a particularly valuable player may be working on a specific task to which he does not contribute much under the specific coalition structure CS. We begin by describing a set of axioms for a value for individual coalitions; that is, suppose that the players have formed the coalition $\mathbf{c} \in [0,1]^n$, what is the amount of payoff that player i should receive from $\mathbf{c}$? Given an OCF game $\mathcal{G} = \langle N, v \rangle$, a *coalition dependent value* is a function Φ whose input is an OCF game $\mathcal{G}$ and a coalition $\mathbf{c} \in [0,1]^n$, and whose output is a payoff vector denoted $\Phi(\mathcal{G}, \mathbf{c})$.

Similarly, a *coalition structure value* (or CS-value when CS is given), is a function Φ whose input is an OCF game $\mathcal{G}$ and a coalition structure $CS \in \mathcal{CS}(N)$, and whose output is a payoff vector $\Phi(\mathcal{G}, CS)$. We now define some axioms for such coalition dependent values, similar to those defined for non-OCF games. We fix a game $\mathcal{G}$ and a coalition $\mathbf{c} \in [0,1]^n$.

Coalitional Efficiency: $\sum_{i=1}^{n} \Phi_i(\mathcal{G}, \mathbf{c}) = v(\mathbf{c})$.

Symmetry: Two players $i, j \in N$ are *symmetric w.r.t.* $\mathbf{c}$ if $v(\mathbf{c}) = v(\mathbf{c}_{i \sim j})$, where $\mathbf{c}_{i \sim j}$ is the coalition $\mathbf{c}$ with the i-th and j-th coordinates exchanged.

Dummy: Set $\mathbf{c}_{-i}$ to be $\mathbf{c}$ with the i-th coordinate set to 0. If $v(\mathbf{c}_{-i}) = v(\mathbf{c})$ then $\Phi_i(N, v, \mathbf{c}) = 0$.

Additivity: Given two OCF games $\mathcal{G} = \langle N, v \rangle$ and $\bar{\mathcal{G}} = \langle N, \bar{v} \rangle$, Φ is additive if $\Phi_i(\mathcal{G}, \mathbf{c}) + \Phi_i(\bar{\mathcal{G}}, \mathbf{c}) = \Phi_i(\mathcal{G} + \bar{\mathcal{G}}, \mathbf{c})$; $\mathcal{G} + \bar{\mathcal{G}} = \langle N, \hat{v} \rangle$, where $\hat{v}(\mathbf{c}) = v(\mathbf{c}) + \bar{v}(\mathbf{c})$.

Given an OCF game $\mathcal{G} = \langle N, v \rangle$ and a coalition $\mathbf{c} \in [0,1]^n$, let us define a non-OCF cooperative game as follows: the player set is N, and the characteristic function $\alpha_{v,\mathbf{c}} : 2^N \to \mathbb{R}$ is defined as $\alpha_{v,\mathbf{c}}(S) = v(\mathbf{c}|_S)$; recall that $\mathbf{c}|_S$ is the coalition $\mathbf{c}$ with all coordinates of $N \setminus S$ set to 0. We define the *coalitional OCF Shapley value* of a game $\mathcal{G}$ and a coalition $\mathbf{c}$ to be the Shapley value of the non-OCF game $\langle N, \alpha_{v,\mathbf{c}} \rangle$.

Theorem 4.16. *The coalitional OCF Shapley value is the only CS-value satisfying the coalitional efficiency, symmetry, dummy and additivity axioms.*

Proof. One can verify that the OCF properties of coalitional efficiency, symmetry, dummy and additivity described above naturally translate to their counterparts in non-OCF games. Hence, the coalitional OCF Shapley value satisfies all properties described above.

To show uniqueness, we use the following construction: given a function $u : 2^N \to \mathbb{R}$, we define an OCF game $\mathcal{G}_u = \langle N, v_u \rangle$ where $v_u : [0,1]^n \to \mathbb{R}$ is as follows: for any $\mathbf{x} \in [0,1]^n$, $v(\mathbf{x}) = u(S)$ if there exists some $S \subseteq N$ such that $\mathbf{x} = \mathbf{c}|_S$ and $v(\mathbf{x}) = 0$ otherwise. We have $\alpha_{v_u,\mathbf{c}}(S) = u(S)$ for any $S \subseteq N$. Therefore, uniqueness of the coalitional OCF Shapley value follows from the uniqueness of the classic Shapley value.

The coalitional OCF Shapley value can be extended to coalition structures by setting $SV_i(\mathcal{G}, CS) = \sum_{\mathbf{c} \in CS} SV_i(\mathcal{G}, \mathbf{c})$. All axioms described above extend naturally from single coalitions to coalition structures. $\square$

The coalitional Shapley value is the only way of dividing coalitional payoffs in a manner that satisfies the four axioms described above; however, these axioms make certain assumptions on the way that players reason about coalition structures. First, the coalitional efficiency axiom requires that all of the value of CS be divided in a coalitionally feasible manner among players. However, this does not require that the coalition structure CS is actually optimal. Second, the symmetry, dummy and additivity axioms all apply to the particular coalition structure CS, rather than to players' overall ability and actual bargaining power.

An alternative approach for measuring power does not assume a preexisting coalition structure, but utilizes a measure of the a-priori marginal contribution of a player, as all players try to maximize social welfare by forming coalition structures. Using a similar characterization to that of Young [1985], we obtain an alternative value. We begin by setting forth the desirable axioms. A value for OCF games is a function Φ whose input is an OCF game $\mathcal{G} = \langle N, v \rangle$, and whose output is a vector in $\mathbb{R}^n$. We describe the following axioms for values for OCF Games:

Strong Monotonicity: a value Φ satisfies *strong monotonicity* if given two OCF games $\mathcal{G} = \langle N, v \rangle$ and $\bar{\mathcal{G}} = \langle N, \bar{v} \rangle$ and some $i \in N$ such that

$$v^*(\mathbf{c}) - v^*(\mathbf{c}_{-i}) \geq \bar{v}^*(\mathbf{c}) - \bar{v}^*(\mathbf{c}_{-i})$$

for all $\mathbf{c} \in [0,1]^n$, then $\Phi_i(\mathcal{G}) \geq \Phi_i(\bar{\mathcal{G}})$.

(2) Symmetry: A value Φ is *symmetric* if for any two symmetric players i, j: $\Phi_i(N, v) = \Phi_j(N, v)$. Two players are symmetric if they are symmetric w.r.t. $\mathbf{c}$ for all $\mathbf{c} \in [0,1]^n$.

(3) Efficiency: A value Φ is *efficient* if $\sum_{i=1}^n \Phi_i(\mathcal{G}) = v^*(N)$.

The strong monotonicity axiom states that if a player i has higher marginal contribution to $v^*(\mathbf{c})$ than to $\bar{v}^*(\mathbf{c})$ for any $\mathbf{c}$, then her value in $\mathcal{G}$ should be higher. This is a generalization of the notion of strong monotonicity given in [Young, 1985] for non-OCF cooperative games. Also note that if two players are symmetric in the OCF sense, then they are symmetric as players in the discrete superadditive cover of the game, U_v.

We define the *OCF Shapley value*, denoted $SV^*(\mathcal{G})$, to be the Shapley value for the discrete superadditive cover of v, $\bar{\mathcal{G}} = \langle N, U_v \rangle$. We begin by showing the following auxiliary lemma.

Lemma 4.17. *The function ρ, which maps an OCF game to its discrete superadditive cover, is onto the set of superadditive classic cooperative games.*

Proof. First, we note that the non-OCF game $\langle N, U_v \rangle$ is superadditive, since the underlying function v^* which is used to define U_v is superadditive. Therefore, the image of ρ is a subset of the class of superadditive non-OCF games. Now, take any superadditive function $u: 2^N \to \mathbb{R}_+$, and consider the function $v_u: [0,1]^n \to \mathbb{R}_+$ defined to be

$$v_u(\mathbf{c}) = \begin{cases} u(S) & \text{if there is some } S \subseteq N \text{ such that } \mathbf{c} = \mathbf{e}^S \\ 0 & \text{otherwise.} \end{cases}$$

Let $U : 2^N \to \mathbb{R}_+$ be the discrete superadditive cover of v_u; for any $S \subseteq N$ we have $U(S) = v_u^*(\mathbf{e}^S) = v_u(S) = u(S)$, so $U \equiv u$. To conclude, for any superadditive non-OCF game $\mathcal{G} = \langle N, u \rangle$, there is some OCF game that is mapped to $\mathcal{G}$ by ρ. □

Theorem 4.18. *$SV^*(\mathcal{G})$ is the only value for OCF games satisfying strong monotonicity, symmetry and efficiency.*

Proof. Strong monotonicity, efficiency and symmetry are inherited from their classic counterparts for $sv(N, U_v)$. Indeed, the sum of the payoffs to the players is $U_v(N) = v^*(N)$, thus efficiency holds. Given two games $\mathcal{G} = \langle N, v \rangle$ and $\bar{\mathcal{G}} = \langle N, \bar{v} \rangle$, suppose that for some $i \in N$ we have that for all $\mathbf{c} \in [0,1]^n$

$$v^*(\mathbf{c}) - v^*(\mathbf{c}_{-i}) \geq \bar{v}^*(\mathbf{c}) - \bar{v}^*(\mathbf{c}_{-i}).$$

In particular we have that for all $S \subseteq N$

$$v^*(\mathbf{e}^S) - v^*(\mathbf{e}^{S\setminus\{i\}}) \geq \bar{v}^*(\mathbf{e}^S) - \bar{v}^*(\mathbf{e}^{S\setminus\{i\}}),$$

which implies that

$$U_v(S) - U_v(S \setminus \{i\}) \geq U_{\bar{v}}(S) - U_{\bar{v}}(S \setminus \{i\}).$$

Therefore, by strong monotonicity of the Shapley value applied to U_v and $U_{\bar{v}}$, we get that strong monotonicity holds for SV^*. Symmetry is shown in a similar manner.

Finally, the Shapely value is unique for the class of superadditive non-OCF games; the set of discrete superadditive covers is the set of superadditive games by Lemma 4.17, thus the OCF Shapley value inherits its uniqueness from the uniqueness of the classic Shapley value. □

The OCF Shapley value is strongly related to the discrete superadditive cover; in fact, using the characterization of the conservative core of an OCF game given in Theorem 2.11, we obtain an even stronger relation between the discrete superadditive cover and the Shapley value, assuming that the discrete superadditive cover is supermodular.

Corollary 4.19. *Given an OCF game $\mathcal{G} = \langle N, v \rangle$ such that the discrete superadditive cover of v is supermodular, then for any optimal coalition structure CS there exists an imputation $\mathbf{x} \in I(CS)$ such that for all $i \in N$, $p_i(CS, \mathbf{x}) = SV_i^*(\mathcal{G})$.*

Proof. Theorem 2.11 shows that for any payoff division $\mathbf{p}$ in the core of the discrete superadditive cover of $\mathcal{G}$ and any optimal coalition structure, there exists an imputation $\mathbf{x} \in I(CS)$ such that $p^i = p_i(CS, \mathbf{x})$ for all $i \in N$. Since U_v is supermodular, the Shapley value of the discrete superadditive cover is a point in the core (see, e.g. Peleg and Sudhölter [2007]). In particular, there is some imputation $\mathbf{x} \in I(CS)$ such that each player receives exactly his Shapley value under $(CS, \mathbf{x})$. □

The two notions of Shapley value for OCF games considered above do not, in general, coincide, even if the coalition structure for which we compute the coalitional OCF Shapley value is socially optimal.

Example 4.20. Let $\mathcal{G}$ be a 3-player threshold task game with player weights w_1, w_2 and w_3 ($N = \{1, 2, 3\}$); we set $w_1 = 5, w_2 = 2, w_3 = 1$. There are two tasks, t_1 and t_2, with the payoff from t_i denoted p_i and the weight required by t_i denoted W_i. We set $p_1 = 6, p_2 = 12$, and $W_1 = 4, W_2 = 8$. Let us compute $SV^*(N, v)$. When player 1 is first or second he has marginal contribution of 6. When he is last, his marginal contribution is 12. Therefore, $SV_1^* = 8$, and, by efficiency and symmetry, $SV_2^* = SV_3^* = 2$. However, consider a coalition structure CS where players work on two copies of t_1, and each of them contributes half of his resources to each copy. That is, player 1 contributes 2.5 to the two copies, player 2 contributes 1, and player 3 contributes 0.5 to each copy. In that case, for any permutation $\sigma \in \Pi(N)$, neither copy of t_1 will be completed by the first two players in the permutation. Therefore, any player has non-zero marginal contribution only if he is last in a permutation, in which case he contributes 12. Therefore all players have an equal Shapley value of $\frac{1}{3} \cdot 12$; i.e.

$$SV_1(\mathcal{G}, CS) = SV_2(\mathcal{G}, CS) = SV_3(\mathcal{G}, CS) = 4.$$

In Example 4.20, player 1 can contribute significantly more than the other players, but his coalitional OCF Shapley value is equal to theirs in CS. This is because in the *specific* coalition structure CS, his marginal contribution is the same as his peers'; if any one of them leaves a coalition, the value of the remaining coalition structure becomes zero.

Note also that if two players in a TTG have $w_i = w_j$, then they are OCF symmetric, which implies that $SV_i^* = SV_j^*$. However, this is not necessarily true for the coalitional Shapley value.

Example 4.21. Consider a 2-player TTG where both players have weight $w \geq 2$, and there are two tasks t_1 and t_2; we again denote the weight of t_i to be W_i and the payoff from t_i to be p_i. We set $W_1 = 2w - 1$ and $W_2 = 1$, and $p_1 = M, p_2 = x$, where $(2w - 1)x \leq M$. We form CS so that player 1 contributes all of her weight to t_1, while player 2 contributes $w - 1$ to t_1, and completes t_2 by herself. When player 1 is first, then $v(\binom{1}{0}) = x$; note that since player 1 has all his weight assigned to a single coalition he is not allowed to optimize and split his weight, completing two copies of t_2 and earning $2x$. On the other hand, player 2 has his weight divided into two tasks under the coalition structure CS, so player 2 can gain $2x$ on her own. This implies that $SV_1(\mathcal{G}, CS) = \frac{1}{2}(x + M + x - 2x) = \frac{M}{2}$, and by efficiency $SV_2(\mathcal{G}, CS) = \frac{M+2x}{2}$.

Example 4.21 implies that the difference between $SV_i^*(N, v)$ and $SV_i(N, v, CS)$ can be arbitrarily large.

Chapter 5

Iterated Revenue Sharing

"And the great owners, who must lose their land in an upheaval, the great owners with access to history, with eyes to read history and to know the great fact: when property accumulates in too few hands it is taken away."
—John Steinbeck, *The Grapes of Wrath*

In previous chapters, we have studied solution concepts for OCF games in a static setting. Players form a coalition structure, divide revenue, and the game ends. In this chapter, we propose a model for iterated revenue sharing, and analyze profit sharing schemes in this setting. Since the model we propose is not based on any well-known solution concepts such as the core, nucleolus or the Shapley value, we begin this chapter with a few motivating examples.

A web routing service deploys various data centers in order to efficiently route data to users; each data center is located in a different location around the globe. A user may download a file from any of the centers, so their download speed depends on the minimal distance to any server. When serving a geographically spread-out population, where a user may download a file from multiple sources, providing a good service to *all* users requires allocating a reasonable amount of bandwidth resources (e.g. appropriate hardware and maintenance services) to all centers. In systems with such complementarities, overall user satisfaction, and ultimately the total revenue the system achieves, can be modeled by a utility function that takes the amount of resources allocated to each component, and outputs the total revenue generated. Such a function could capture the notion that the resources are not complete substitutes for one another. In this setting, money plays two roles: first, it is actual system revenue; second, it is a *resource*, used as a proxy for generating future revenue. In other words, the web routing service's revenue at time t serves as its available resources at time $t + 1$.

While the system as a whole benefits from allocating revenue among several data centers, individual data centers do not have the global objective of overall productivity in mind. Each data center simply wishes to maximize the share it receives, increasing its own welfare. Indeed, a data center would push for a large share of the profits, even if this leads to lower overall revenue: a larger share of a smaller pie may be better than a smaller share of a larger pie. However, since data centers are mutually dependent to a great extent, no individual center would want to receive an unduly high share of the profits, as this could decrease overall revenue to such a degree that future (individual) profits are worse. To conclude, when negotiating a share of the budget, individual data centers want to increase their own share of the budget allocated, but would not argue for an excessively large portion, as this may hurt their long-term welfare. This setting raises several interesting questions. First, can one compute a budget division that maximizes long-term revenue? Will individual data centers push for a higher share at the expense of social welfare? Can the objectives of the individual data centers and the company as a whole be reconciled?

This type of interplay between individual desires and global utility is common in settings where budgets need to be divided. Consider, for example, a large software company with several departments. Each department works on a specific product and is allocated a budget at the end of some time period (say, biannually). Each department uses its budget to produce a product (a tablet, a gaming console, an operating system etc.), which is pooled at the end of the "fiscal round" and redistributed. Here, as is the case with the previous example, money plays a dual role: when assigned to the divisions, money is a resource, who in turn use it in order to generate money as revenue. We assume that there are complementarities between divisions, i.e., profits are increased by investing in all divisions; this may be due to the fact that different products appeal to different markets —it is better to participate in several markets, rather than focusing solely on OS development— or because some divisions, such as IT and HR, provide services to other divisions. In fact, no division in a company can truly function as an autonomous unit: each division requires the help of others in order to operate. Divisions are faced with two conflicting agendas. On the one hand, each division wants to maximize the revenue share that it receives (a division with higher revenue can increase employee salaries, hire more employees to decrease workload on others, invest in better equipment and software, etc.); on the other hand, no division wants to receive the entire share of the revenue, as this may hurt future profits and result in lower future revenue for the division. If one division receives a disproportionately large share of the profits, total company revenue may dramatically decrease, which, in turn, hurts future revenue. The same questions asked in the trading example arise here: what is the socially optimal way of dividing revenue? What is the revenue sharing scheme that is best for an individual division? Is there a way to divide profits such that individual divisions are happy and social welfare is high?

The model we consider in this chapter is similar in its essence to the OCF model previously described. Just like the OCF model, each player has a resource, and a given characteristic function describes the revenue that is generated by players allocating resources to a collaborative task. We propose two major variants to the OCF model. The first is the structure of player incentives; instead of wanting to maximize only immediate revenue, players in this setting also care about future revenue, thus making them more inclined to help others if that results in higher future revenue for themselves. Second, we do not consider the formation of coalition structures, or alternatively, we allow coalition structures but do not impose coalitional efficiency. In other words, once revenue is generated, there are no constraints on how it can be divided among the players.

We do not employ the mechanics of arbitration functions in this chapter. This is because iterated revenue sharing requires a different model of incentives than that used in the static setting. As we later describe, players do not deviate from outcomes in this setting, but rather use a measure of regret to assess the long-term desirability of a payoff division.

We are given a valuation function $v : \mathbb{R}^n_+ \to \mathbb{R}$ and an initial endowments vector $\mathbf{w}(0) \in \mathbb{R}^n$, where $w_i(0)$ is the initial amount of resources (money) that player i possesses. Our objective is to find a sequence of resource allocations that maximize long-term social welfare on the one hand, and on the other hand, are such that players do not want to increase their share.

We assume that players assess their welfare by comparing it to the amount of money they would have received under some other sequence of revenue allocations. That is, the goal of players in this setting is to minimize their regret: the difference between what they are getting at current, and what they could have received under a more agreeable revenue allocation sequence.

Our results focus on stationary contracts, in which each player receives a constant share of the profits at each round (e.g. player 1 receives 60% of the profits, while player 2 receives 40% of the profits at every round). When limited to this class of contracts, we show when individual incentives align with group incentives in the limit (Section 5.2). In other words, under certain conditions, while it is always the case that players want to receive a higher share of the profits than what they are receiving under a socially optimal revenue division, the difference between what a player actually gets in a socially optimal allocation and what a player *wants* to get goes to 0 as t approaches infinity.

One of the immediate corollaries of our results is that when regret goes to zero, a sufficiently patient player would be incentivized to truthfully report his private information, since that would enable the center to correctly choose the socially optimal revenue division, which is also what is optimal for the individual player in the long run. In slightly more detail, suppose that in order to decide on a payoff division, each player needs to report to a center some private information. The actual value of this private information is used when computing v, while the reported information is used in order to decide on a payoff division. For example, in a network flow game where players are edges, it might be the case that players need to report their edge capacities. If a player's regret goes to 0 as t increases, then he will not be incentivized to misreport his private information.

In Section 5.3, we study the effects of discounted returns on our model. We show that when discounted rewards are in place, starting conditions (e.g., the initial resource vector used) can greatly affect the regret players experience; hoewever, in many cases, adding discounted rewards does not much affect the results in Section 5.2.

Finally, in Section 5.4, we show some applications of our results to well-known economic functions, namely CES, Cobb-Douglas and Leontief production functions. We also study network flow games and compare our payoff division with the core payoff allocation.

5.1 Preliminaries

We assume that players are interested in maximizing their *long-term* profits; for example, suppose that we have a two-player game whose valuation function is given by $v(x,y) = xy$, and players start with 10 units of resource each. This can be written as a characteristic function game where $v(\{1\}) = v(\{2\}) = 0$ and $v(\{1,2\}) = 100$. A quick analysis of this game would show that any payoff division is in the core of this game. However, if one assumes that player payoffs at each round are their resources for the next, not all core allocations are equally good for ensuring high future returns. For example, if player 1 demands all of the profits in the first round, leaving player 2 with 0, then the amount of revenue to be divided in the next round is $v(100,0) = 0$. In fact, the payoff division that ensures maximal profits for player 1 in the limit is *equal sharing*, i.e., giving 50 to both players (this will be evident in the analysis we present in Section 5.2).

In this analysis we do not assume the formation of overlapping coalition structures. In particular, we do not use v^* rather than v in our optimization goals. However, overlapping coalition structures are not unreasonable to assume in this setting. Our results do apply if one uses v^* instead of v for the purpose of computing the amount of revenue that is to be shared; however, they only apply if coalitional efficiency is not enforced, i.e. if we do not require that payoffs from a given partial coalition $\mathbf{c} \in CS$ are given only to $supp(\mathbf{c})$. Enforcing coalitional efficiency leads to a much more complicated dynamic that is left for future work.

In general, the player set N starts with an initial endowments vector $\mathbf{w}(0)$ and is given a valuation function $v: \mathbb{R}^n_+ \to \mathbb{R}$. Players go through the following process: initially, players select a vector $\mathbf{x} \in \Delta_n$, where $\Delta_n = \{\mathbf{x} \in \mathbb{R}^n_+ \mid \sum_{i=1}^n x_i = 1\}$. The revenue at time 1 is $v(\mathbf{w}(0))$, which we denote V_1; for every time step $t > 1$, let V_t be defined as $v(x_1 \cdot V_{t-1}, \ldots, x_n \cdot V_{t-1})$. In other words, given the initial profit sharing agreement $\mathbf{x}$, the resources available to player i at time t are given by $x_i V_{t-1}$, the amount of money allocated to him based on $\mathbf{x}$; the revenue is then determined by evaluating v on the vector $(x_1 \cdot V_{t-1}, \ldots, x_n \cdot V_{t-1})$. We call $\mathbf{x} \in \Delta_n$ a *stationary contract*.

Given a stationary contract $\mathbf{x} \in \Delta_n$, the *total welfare at time* t is $sw_t(\mathbf{x}) = \sum_{h=1}^t v_h(\mathbf{x})$. We say that a stationary contract $\mathbf{x}^*$ is *optimal at time* t if $\mathbf{x}^*$ maximizes $sw_t(\mathbf{x})$. We say that a contract $\mathbf{x}^*$ is *universally optimal* if $\mathbf{x}^*$ is optimal at time t for all $t \in \mathbb{N}$. We say that a contract is *pointwise optimal at time* t if $\mathbf{x} \in \arg\max_{\mathbf{x} \in \Delta_n} v_t(\mathbf{x})$, and that it is *universally pointwise optimal* if it is pointwise optimal for all $t \in \mathbb{N}$. We note that if $\mathbf{x}^*$ is universally pointwise optimal then it is universally optimal.

We assume that players are not interested in the global social welfare provided by a contract, but rather in their own revenue. Given a player $i \in N$, the benefit that player i receives from the contract $\mathbf{x}$ at time t is given by the function $u_{i,t}(\mathbf{x}) = x_i^t v_t(\mathbf{x})$; we also refer to $u_{i,t}(\mathbf{x})$ as player i's *share* at time t under $\mathbf{x}$. We write $U_{i,t}(\mathbf{x}) = \sum_{h=1}^t u_{i,t}(\mathbf{x})$. We say that $\mathbf{x}$ is *individually optimal for player i at time t* if $\mathbf{x}$ maximizes $U_{i,t}$ over Δ_n, *universally individually optimal* if $\mathbf{x}$ maximizes $U_{i,t}$ for all $t \in \mathbb{N}$, *pointwise individually optimal* if $\mathbf{x}$ maximizes $u_{i,t}$, and *universally pointwise individually optimal* if $\mathbf{x}$ maximizes $u_{i,t}$ for all $t \in \mathbb{N}$. Again, note that universally pointwise individually optimality implies the other properties.

Finally, given a player $i \in N$ and a contract $\mathbf{x} \in \Delta_n$, let $p\text{-}reg_i(\mathbf{x}, t) = \max_{\mathbf{y} \in \Delta_n} u_{i,t}(\mathbf{y}) - u_{i,t}(\mathbf{x})$ be his *pointwise regret at time* t. Intuitively, the desirability of a contract at time t is measured against the revenue that player i could have secured under a more favorable contract.

5.1.1 Accounting for Individual Utilities

By taking the share that player i receives as his utility, we are assuming that player i wants to receive a higher share of the profits at every time step. Without an explicit notation for player i's utility for the amount of money he receives, it is not a-priori clear why we set player 1's objective to be maximizing profits. Moreover, we assume that at every round, all players invest all of their resources back into generating profits.

We can easily modify our setting so as to accommodate for a fixed percentage of player profits going towards individual utility. We can achieve this by adding a parameter c_i for every $i \in N$, where $c_i \in [0, 1]$ is the share of the profits that i keeps for himself; c_i can represent labor costs, server costs or employee benefits; all of our results can be modified accordingly. Assuming that the share of the profits that goes towards individual utility is fixed is perfectly acceptable in settings where players represent bodies with no ability to spend money outside the context of the production function (e.g. stock traders in a firm, university departments or divisions in a company).

We implicitly assume that there exists some utility function for every player, without ever referring to it; in other words, we assume that players prefer more money to less, without specifying an exact reason for this. Our model can be extended to more complex notions of utility. By adding utility functions, we can assume that each player wants to maximize his utility by allocating resources to personal use. At each round, every player needs to choose the amount of money that he allocates to individual use, and the amount he wants to "reinvest" in producing future returns. A greedy player would choose to allocate most (or all) of his initial revenue in personal happiness, losing on possibly high future revenue; an overly cautious player would allocate most of his revenue towards future rewards, which results in consistently high future rewards with little individual day to day happiness.

A setting where players can choose the share of their revenue that is reinvested into the "public function" at each round is a promising extension of our work. In this setting, at every time step t, every player needs to choose a share $c_i(t)$ that is kept by the player, while the rest of the profit is reinvested. Note that if we assume that players are completely myopic (i.e. they only care about the profits at time 1), then $c_i(1) = 1$ for all $i \in N$, and the framework we describe collapses to that of OCF games. Non-myopic players, on the other hand, need to choose a sequence $(c_i(t))_{t=1}^{\infty}$ that maximizes their individual utilities. Studying the Nash equilibria of the resulting game would be an interesting direction for future work, as it would combine a notion of Nash equilibrium and cooperative equilibrium in a non-trivial manner.

5.2 Optimization and Regret in Homogeneous Functions

In order to have a robust model of cooperation and individual incentives, we wish to capture some notion of complementarity among players. If players are actually better off without allocating resources to some subset of players, then there is little a-priori incentive to collaborating with them. This notion is captured via the idea of *mutual dependency*.

Definition 5.1. A function $v : \mathbb{R}^n \to \mathbb{R}$ satisfies mutual dependency if for all $\mathbf{x} \in \Delta_n$ and all $i \in N$, if $x_i = 0$, then $v(\mathbf{x}) = 0$.

We note that if one assumes that v^* is used rather than v (i.e. players are allowed to form overlapping coalition structures), then it is reasonable to assume that mutual dependency holds: v^* makes optimal use of player resources, ensuring that if a player can contribute something, then he will.

Of course, mutual dependency does not always hold; as the following example shows: games where the only thing determining the worth of a player is the amount of money he possesses —such as TTGs— do not satisfy the mutual dependency property.

Example 5.2. Consider a setting where for any $\mathbf{x} \in \mathbb{R}^n_+$ we have that $v(\mathbf{x}) = f(\sum_{i=1}^n x_i)$, where $f : \mathbb{R}_+ \to \mathbb{R}_+$ is a monotone function. In that case, the value of v at time t is simply $f^t(W)$, where $W = \sum_{i \in N} w_i(0)$, and f^t is the composition of f with itself t times. In this setting, the only thing determining the worth of a player is the amount of money he brings to the table at time t. Mutual dependency does not hold here. In this setting, any choice $\mathbf{x} \in \Delta_n$ is socially optimal (both universally and pointwise).

We can assume that mutual dependency does not hold; the main reason why mutual dependency is required is since we need contracts that maximize welfare or individual utility are in the interior of Δ_n.

We start by showing a negative result, which, despite its simplicity, motivates the reasoning behind the rest of our results. Given an optimal (or pointwise optimal) contract $\mathbf{x}^* \in \Delta_n$, can it be the case that $\mathbf{x}^*$ is also individually optimal for some player $i \in N$? The following lemma answers this question in the negative.

Lemma 5.3. *Suppose that v is differentiable, and let $F : \mathbb{R}^n \to \mathbb{R}$ be in $\{v_t, sw_t\}$; Given some $\mathbf{x}^*$ that is a global maximum of F in the interior of Δ_n, then 0 is not a maximum of the function $g_{ij}(x) = (x_i^* + x)F(\mathbf{y}(x))$, where $\mathbf{y}(x)$ equals $\mathbf{x}^*$ on all coordinates but i, j, and $y_i(x) = x_i^* + x, y_j(x) = x_j^* - x$.*

Proof. First, observe that if v is differentiable, then F is differentiable as well. We note that if $\mathbf{x}^* \in \Delta_n$ is a global maximum of F in the interior of Δ_n, then $\frac{\partial F}{\partial x_i}(\mathbf{x}^*) = \frac{\partial F}{\partial x_j}(\mathbf{x}^*)$ for all $i, j \in N$. Taking the derivative of g_{ij}, we get $\frac{\partial g_{ij}}{\partial x}(x) =$

$F(\mathbf{y}(x)) + (x_i^* + x)(\frac{\partial F}{\partial x_j}(\mathbf{y}(x)) - \frac{\partial F}{\partial x_i}(\mathbf{y}(x)))$; Thus $\frac{\partial g_{ij}}{\partial x}(0) = F(\mathbf{x}^*) > 0$. In particular 0 is not a maximum of g_{ij}, which concludes the proof. □

The proof of Lemma 5.3 is simple, but we wish to stress its importance. Lemma 5.3 implies that when a contract is socially optimal, it is necessarily suboptimal for all players. This implies that social welfare and individual gains are always at odds. Can we, under certain conditions, mitigate this effect? In other words, given an optimal stationary contract $\mathbf{x}^* \in \Delta_n$, can we guarantee that the regret of players goes to 0 in the limit? In what follows, we show when this is indeed the case, if the production function v is homogeneous of degree $k \geq 1$ and has a unique maximum over Δ_n (e.g. if v is concave). Assuming that we allow players to optimize and form coalition structures, we need to assume that v^* is homogeneous rather than v. As we have shown in Lemma 2.25, if v is k-homogeneous then so is v^*; if v is 1-homogeneous, other properties are easily derived.

Lemma 5.4. *If v is 1-homogeneous, then v^* is concave.*

Proof. Since v^* is superadditive, we know that $v^*(\alpha \mathbf{x} + (1-\alpha)\mathbf{y}) \geq v^*(\alpha \mathbf{x}) + v^*((1-\alpha)\mathbf{y})$; since v is 1-homogeneous, then so is v^*, implying that $v^*(\alpha \mathbf{x} + (1-\alpha)\mathbf{y}) \geq \alpha v^*(\mathbf{x}) + (1-\alpha)v^*(\mathbf{y})$. □

As shown in Lemma 2.25, when v^* is k-homogeneous of degree $k \geq 1$, the refined core is not empty. The results in this section show that other highly desirable behaviors occur when v is homogeneous of degree $k \geq 1$, i.e. exhibiting increasing returns to scale.

We begin by deriving a closed-form formula for $v_t(\mathbf{x})$, given that players commit to the stationary contract $\mathbf{x}$, when v is homogeneous.

Lemma 5.5. *If v is homogeneous of degree k, then $v_t(\mathbf{x}) = V_1^{k^{t-1}} v(x)^{\sum_{h=0}^{t-2} k^h}$ for all $t \geq 2$, where $V_1 = v(\mathbf{w}(0))$ denotes the welfare at time 1.*

Proof. Our proof is by induction on t. For $t = 2$ we have

$$v_2(\mathbf{x}) = v(v(\mathbf{w}(0))\mathbf{x}) = v(\mathbf{w}(0))^k v(\mathbf{x}) = V_1^{k^1} v(\mathbf{x})^{k^0}.$$

Now, assuming that the claim holds for $t-1$, we show it holds for t:

$$\begin{aligned} v_t(\mathbf{x}) &= v(v_{t-1}(\mathbf{x})\mathbf{x}) = v_{t-1}(\mathbf{x})^k v(\mathbf{x}) \\ &\stackrel{\text{i.h.}}{=} \left(V_1^{k^{t-2}} v(\mathbf{x})^{\sum_{h=0}^{t-3} k^h}\right)^k v(\mathbf{x}) \\ &= V_1^{k^{t-1}} v(\mathbf{x})^{\sum_{h=1}^{t-2} k^h} v(\mathbf{x}) = V_1^{k^{t-1}} v(x)^{\sum_{h=0}^{t-2} k^h} \end{aligned}$$

□

One immediate corollary of Lemma 5.5 is that if v is homogeneous, then a contract that maximizes v is an optimal stationary contract for v_t for all $t \in \mathbb{N}$.

Corollary 5.6. *If v is homogeneous of degree k and $\mathbf{x}^* \in \arg\max v$ over Δ_n, then $\mathbf{x}^*$ is a universally optimal and pointwise optimal stationary contract.*

Proof. Let us write $F_t(x) = V_1^{k^{t-1}} x^{H_t}$, where $H_t = \sum_{h=0}^{t-2} k^h$; F_t is a strictly monotone increasing function of x when $x \geq 0$. If $\mathbf{x}^* \in \arg\max v$ then for all $\mathbf{y} \in \Delta_n$ we have that $v(\mathbf{x}^*) \geq v(\mathbf{y})$, so $F_t(v(\mathbf{x}^*)) \geq F_t(v(\mathbf{y}))$ for all $\mathbf{y} \in \Delta_n$, hence $\mathbf{x}^* \in \arg\max F_t \circ v = v_t$, which implies that $\mathbf{x}^*$ is pointwise optimal at time t for all t. Similarly, let us write $G_t(x) = \sum_{s=1}^{t} F_s(x)$; then $G_t(v(\mathbf{x})) = \mathrm{sw}_t(\mathbf{x})$, and is again a monotone increasing function of x, and if $\mathbf{x}^* \in \arg\max_{\mathbf{x} \in \Delta_n} v(\mathbf{x})$, then $\mathbf{x}^*$ maximizes $G_t \circ v = \mathrm{sw}_t$ over Δ_n. Note that the other direction holds as well: if $\mathbf{x}^*$ maximizes $v_t(\mathbf{x})$ over Δ_n, then $\mathbf{x}^*$ maximizes $v(\mathbf{x})$ over Δ_n. □

Before we proceed, we state the following properties of homogeneous functions without proof.

Proposition 5.7. *Let $f : \mathbb{R}^n \to \mathbb{R}$ be a k-homogeneous, differentiable function.*

1. *Let $F(\mathbf{x}) = f(\mathbf{x})^m$; then F is mk-homogeneous.*

2. *Let $G(\mathbf{x}) = cf(\mathbf{x})$ for some constant $c \neq 0$; then G is k-homogeneous.*

3. *Let $H_i(\mathbf{x}) = x_i f(\mathbf{x})$; then H_i is $(k+1)$-homogeneous.*

4. *The point $\mathbf{x}^*$ is a critical point of f in the interior Δ_n if and only if $\frac{\partial f}{\partial x_i}(\mathbf{x}^*) = kf(\mathbf{x}^*)$ for all $i \in N$.*

Now, let us observe some properties of sequences of individually pointwise optimal stationary contracts for player i.

Lemma 5.8. *Suppose that v is k homogeneous. Let $\mathbf{x}_t^*$ be a point in Δ_n such that $\mathbf{x}_t^* \in \arg\max_{\mathbf{x} \in \Delta_n} u_{i,t}(\mathbf{x})$, and $\mathbf{x}^*$ be a point in Δ_n such that $\mathbf{x}^* \in \arg\max_{\mathbf{x} \in \Delta_n} v(\mathbf{x})$; then $x_{i,t}^* \geq x_i^*$.*

Proof. Let $\mathbf{x}^* \in \arg\max_{\mathbf{x} \in \Delta_n} v(\mathbf{x})$; by Corollary 5.6, for all $t \in \mathbb{N}$, we have $\mathbf{x}^* \in \arg\max_{\mathbf{x} \in \Delta_n} v_t(\mathbf{x})$. Assume that the claim does not hold; then $x_{i,t}^* < x_i^*$. Therefore,

$$u_{i,t}(\mathbf{x}_t^*) = x_{i,t}^* v_t(\mathbf{x}_t^*) \leq x_{i,t}^* v_t(\mathbf{x}^*) < x_i^* v_t(\mathbf{x}^*).$$

We get that $u_{i,t}(\mathbf{x}_t^*) < x_i^* v_t(\mathbf{x}^*) = u_{i,t}(\mathbf{x}^*)$. Therefore $\mathbf{x}_t^* \notin \arg\max_{\mathbf{x} \in \Delta_n} u_{i,t}(\mathbf{x})$, a contradiction. □

Since Lemma 5.8 holds for any $t \in \mathbb{N}$, we get as a corollary that there is some constant $c > 0$ such that $x_{i,t}^* > c$ for all $t \in \mathbb{N}$; therefore, for all $t \in \mathbb{N}$, we have that $u_{i,t}(\mathbf{x}_t^*) \geq cv_t(\mathbf{x}^*)$, where $\mathbf{x}_t^*$ is individually pointwise optimal for i, and $\mathbf{x}^*$ is universally optimal.

We now show an important first step in establishing the connection between the degree of homogeneity and individual regret.

Theorem 5.9. *Suppose that $v : \mathbb{R}^n \to \mathbb{R}$ is k-homogeneous and differentiable; given some $i \in N$, let $(\mathbf{x}_t^*)_{t=1}^{\infty}$ be a sequence of individually pointwise optimal stationary contracts for i, such that $\lim_{t \to \infty} \mathbf{x}_t^* = \mathbf{x}^*$, where $\mathbf{x}^*$. The point $\mathbf{x}^*$ is a critical point of v over Δ_n if and only if $k \geq 1$.*

Proof. We note that because of mutual dependency, $\mathbf{x}_t^*$ are interior points of Δ_n. By Proposition 5.7, if $\mathbf{x}_t^*$ is in $\arg\max_{\mathbf{x}\in\Delta_n} u_{i,t}(\mathbf{x})$, then $\frac{\partial u_{i,t}}{\partial x_j}(\mathbf{x}_t^*) = (H_t k + 1)u_{i,t}(\mathbf{x}_t^*)$ for all $j \in N$. For $j \neq i$ we get

$$\frac{\partial u_{i,t}}{\partial x_j}(\mathbf{x}_t^*) = x_{i,t}^* V_1^{k^{t-1}} H_t v(\mathbf{x}_t^*)^{H_t - 1} \frac{\partial v}{\partial x_j}(\mathbf{x}_t^*)$$

$$= H_t \frac{1}{v(\mathbf{x}_t^*)} \frac{\partial v}{\partial x_j}(\mathbf{x}_t^*) u_{i,t}(\mathbf{x}_t^*),$$

so for all $j \neq i$ we have

$$H_t \frac{1}{v(\mathbf{x}_t^*)} \frac{\partial v}{\partial x_j}(\mathbf{x}_t^*) u_{i,t}(\mathbf{x}_t^*) = (H_t k + 1) u_{i,t}(\mathbf{x}_t^*),$$

or that

$$\frac{\partial v}{\partial x_j}(\mathbf{x}_t^*) = (k + \frac{1}{H_t}) v(\mathbf{x}_t^*).$$

A similar computation for player i gives:

$$\frac{\partial u_{i,t}}{\partial x_i}(\mathbf{x}_t^*) = v_t(\mathbf{x}_t^*) + x_{i,t}^* \frac{\partial v_t}{\partial x_i}(\mathbf{x}_t^*)$$

$$= v_t(\mathbf{x}_t^*) + x_{i,t}^* V_1^{k^{t-1}} H_t v(\mathbf{x}_t^*)^{H_t - 1} \frac{\partial v}{\partial x_i}(\mathbf{x}_t^*)$$

$$= v_t(\mathbf{x}_t^*) + x_{i,t}^* H_t v_t(\mathbf{x}_t^*) \frac{1}{v(\mathbf{x}_t^*)} \frac{\partial v}{\partial x_i}(\mathbf{x}_t^*)$$

$$= v_t(\mathbf{x}_t^*) \left(1 + x_{i,t}^* H_t \frac{1}{v(\mathbf{x}_t^*)} \frac{\partial v}{\partial x_i}(\mathbf{x}_t^*)\right)$$

Combining this with Proposition 5.7, we obtain that

$$v_t(\mathbf{x}_t^*) \left(1 + x_{i,t}^* H_t \frac{1}{v(\mathbf{x}_t^*)} \frac{\partial v}{\partial x_i}(\mathbf{x}_t^*)\right) = u_{i,t}(\mathbf{x}_t^*),$$

which simplifies to

$$\frac{\partial v}{\partial x_i}(\mathbf{x}_t^*) = (k + \frac{1}{H_t}) v(\mathbf{x}_t^*) - \frac{v(\mathbf{x}_t^*)}{x_{i,t}^* H_t},$$

We are now ready to show the relation between the degree of k and the convergence of $\mathbf{x}_t^*$. Recall that $H_t = \sum_{h=0}^{t-2} k^h$; if $k \geq 1$ then $\lim_{t \to \infty} H_t = \infty$, otherwise $\lim_{t \to \infty} H_t = \frac{1}{1-k}$. Therefore, if $k \geq 1$ then for all $j \neq i$

$$\frac{\partial v}{\partial x_j}(\mathbf{x}^*) = \lim_{t \to \infty} \frac{\partial v}{\partial x_j}(\mathbf{x}_t^*) = kv(\mathbf{x}^*).$$

By Lemma 5.8, there is some constant $c > 0$ such that for all $t \in \mathbb{N}$, $x_{i,t}^* \geq c$; noting that

$$(k + \frac{1}{H_t}) v(\mathbf{x}_t^*) - \frac{\max_{\mathbf{x} \in \Delta_n} v(\mathbf{x})}{cH_t} \leq \frac{\partial v}{\partial x_i}(\mathbf{x}_t^*) \leq (k + \frac{1}{H_t}) v(\mathbf{x}_t^*)$$

we get that
$$\frac{\partial v}{\partial x_i}(\mathbf{x}^*) = \lim_{t \to \infty} \frac{\partial v}{\partial x_i}(\mathbf{x}_t^*) = kv(\mathbf{x}^*).$$

According the Proposition 5.7, since $\frac{\partial v}{\partial x_j}(\mathbf{x}^*) = kv(\mathbf{x}^*)$ for all $j \in N$, $\mathbf{x}^*$ is a critical point of v over Δ_n, which concludes the proof for the case $k \geq 1$.

Suppose that $k < 1$; since $\lim_{t \to \infty} H_t = \frac{1}{1-k}$, for all $j \neq i$,
$$\frac{\partial v}{\partial x_j}(\mathbf{x}^*) = \lim_{t \to \infty} \frac{\partial v}{\partial x_j}(\mathbf{x}_t^*) = (k+1-k)v(\mathbf{x}_t^*) = v(\mathbf{x}^*).$$

On the other hand, for i, taking limits, we have
$$\frac{\partial v}{\partial x_i}(\mathbf{x}^*) = v(\mathbf{x}^*) - (1-k)\frac{v(\mathbf{x}^*)}{x_i^*} = v(\mathbf{x}^*)(1 - \frac{1-k}{x_i^*}).$$

According to Proposition 5.7, if $\mathbf{x}^* \in \arg\max_{\mathbf{x} \in \Delta_n} v(\mathbf{x})$ then $\frac{\partial v}{\partial x_i}(\mathbf{x}^*) = \frac{\partial v}{\partial x_j}(\mathbf{x}^*)$ for all $j \neq i$. However, if $\frac{\partial v}{\partial x_i}(\mathbf{x}^*) = \frac{\partial v}{\partial x_j}(\mathbf{x}^*)$ then $\frac{1-k}{x_i^*} = 0$, or $k = 1$, which is a contradiction to the assumption that $k < 1$. □

We observe that if v satisfies mutual dependency, then any stationary contract $\mathbf{x}^* \in \Delta_n$ that is universally optimal must be in the interior of Δ_n. We can show that convergence to the optimal stationary contract occurs in continuous, homogeneous functions whenever the optimal contract is a unique critical point in the interior of Δ_n. We first show the following lemma

Lemma 5.10. *Let D be a convex and compact subset of $\mathbb{R}^n$. Suppose that $f : D \to \mathbb{R}^m$ has a unique point $\mathbf{x}^* \in D$ such that $f(\mathbf{x}^*) = \mathbf{c}^*$, and suppose that $\lim_{t \to \infty} f(\mathbf{x}_t) = \mathbf{c}^*$, then $\lim_{t \to \infty} \mathbf{x}_t = \mathbf{x}^*$.*

Proof. Assume otherwise; then in particular, there exists a subsequence of $\mathbf{x}_t$, $\mathbf{x}_{t_\ell}$ so that for all $\ell \in \mathbb{N}$ we have $|\mathbf{x}_{t_\ell} - \mathbf{x}^*| > \varepsilon_0$ for some fixed $\varepsilon_0 > 0$. Since D is compact, there is some subsequence of $\mathbf{x}_{t_\ell}$, say $\mathbf{x}_{t_{\ell,k}}$ that converges to some point $\mathbf{y}^*$, which, by assumption is not $\mathbf{x}^*$. On the one hand, we have $\lim_{k \to \infty} f(\mathbf{x}_{t_{\ell,k}}) = f(\mathbf{y}^*)$; on the other hand, since $\mathbf{x}_{t_{\ell,k}}$ is a subsequence of $\mathbf{x}_t$, we have $\lim_{k \to \infty} f(\mathbf{x}_{t_{\ell,k}}) = f(\mathbf{x}^*)$, thus it must be the case that $f(\mathbf{x}^*) = f(\mathbf{y}^*) = \mathbf{c}^*$, a contradiction to the fact that $\mathbf{x}^*$ is the only point in D such that $f(\mathbf{x}) = \mathbf{c}^*$. □

Using Lemma 5.10, we can now prove the following corollary.

Corollary 5.11. *If $v : \mathbb{R}^n \to \mathbb{R}$ is differentiable and homogeneous of degree $k \geq 1$, such that v has a unique critical point $\mathbf{x}^*$ in the interior of Δ_n, then any sequence $\mathbf{x}_t \in \arg\max_{\mathbf{x} \in \Delta_n} u_{i,t}(\mathbf{x})$ of contracts in the interior of Δ_n converges to $\mathbf{x}^*$, and $\mathbf{x}^*$ is a global maximum of v over Δ_n.*

Proof. Let us define the mapping $G_v : \Delta_n \to \mathbb{R}$ as
$$G_v(\mathbf{x}) = \max_{j,k \in N}(|\frac{\partial v}{\partial x_j}(\mathbf{x}) - \frac{\partial v}{\partial x_k}(\mathbf{x})|).$$

As observed in Theorem 5.9, $\frac{\partial v}{\partial x_j}(\mathbf{x}_t^*) = \frac{\partial v}{\partial x_k}(\mathbf{x}_t^*) = (k + \frac{1}{H_t})v(\mathbf{x}_t^*)$ for all $j, k \neq i$. Moreover, for all $j \neq i$, $|\frac{\partial v}{\partial x_j}(\mathbf{x}_t^*) - \frac{\partial v}{\partial x_i}(\mathbf{x}_t^*)| = |\frac{v(\mathbf{x}_t^*)}{x_{i,t}^* H_t}|$. Since $k \geq 1$, $\lim_{t \to \infty} G_v(\mathbf{x}) = \lim_{t \to \infty} |\frac{v(\mathbf{x}_t^*)}{x_{i,t}^* H_t}| = 0 = G_v(\mathbf{x}^*)$.

Since v has a unique critical point over the interior of Δ_n, and all points in $\mathbf{x}_t^*$ are at a distance of at least some constant $c > 0$ from the boundary of Δ_n by Lemma 5.8, it must be, by Lemma 5.10 that $\lim_{t \to \infty} \mathbf{x}_t^* = \mathbf{x}^*$. Finally, since v is continuous, it must have a maximum over Δ_n; since this maximum does not occur on the boundary of Δ_n, it must be that $\mathbf{x}^*$ is the maximum of v over Δ_n. In other words, in the case of a single critical point, all sequences of individually optimal contracts converge to the optimal contract. □

Corollary 5.11 means that if v has a unique critical point over Δ_n (e.g. if v is a strictly concave function), and is homogeneous of degree $k \geq 1$, then any sequence of individually optimal contracts for player i will converge to the universally optimal contract $\mathbf{x}^*$. Combining these results we obtain that

Theorem 5.12. *Suppose that the following conditions on v hold:*

1. *v satisfies mutual dependency.*
2. *v is homogeneous of degree $k \geq 1$.*
3. *v is differentiable over Δ_n.*
4. *v has a unique critical point over the interior of Δ_n.*

Then for any $i \in N$, $\lim_{t \to \infty} p\text{-}reg_i(\mathbf{x}^*, t) = 0$, where $\mathbf{x}^* \in \arg\max_{\mathbf{x} \in \Delta_n} v(\mathbf{x})$; moreover, if $\mathbf{x}_t^*$ is a sequence of pointwise individually optimal stationary contracts for player i, then $\lim_{t \to \infty} \mathbf{x}_t^* = \mathbf{x}^*$.

Proof. We wish to show that if $v : \mathbb{R}^n \to \mathbb{R}$ is differentiable and homogeneous of degree $k \geq 1$, such that v has a unique critical point $\mathbf{x}^*$ in the interior of Δ_n, then any sequence $\mathbf{x}_t \in \arg\max_{\mathbf{x} \in \Delta_n} u_{i,t}(\mathbf{x})$ of contracts converges to $\mathbf{x}^*$, and $\mathbf{x}^*$ is a maximum of v over Δ_n.

Let us define the mapping $G_v : \Delta_n \to \mathbb{R}$ as $G_v(\mathbf{x}) = \max_{j,k \in N}(|\frac{\partial v}{\partial x_j}(\mathbf{x}) - \frac{\partial v}{\partial x_k}(\mathbf{x})|)$. As observed in Theorem 5.9, $\frac{\partial v}{\partial x_j}(\mathbf{x}_t^*) = \frac{\partial v}{\partial x_k}(\mathbf{x}_t^*) = (k + \frac{1}{H_t})v(\mathbf{x}_t^*)$ for all $j, k \neq i$. Moreover, for all $j \neq i$, $|\frac{\partial v}{\partial x_j}(\mathbf{x}_t^*) - \frac{\partial v}{\partial x_i}(\mathbf{x}_t^*)| = |\frac{v(\mathbf{x}_t^*)}{x_{i,t}^* H_t}|$. Since $k \geq 1$, $\lim_{t \to \infty} G_v(\mathbf{x}) = \lim_{t \to \infty} |\frac{v(\mathbf{x}_t^*)}{x_{i,t}^* H_t}| = 0 = G_v(\mathbf{x}^*)$. □

The unique critical point assumption is essential to the proof of Theorem 5.12, as the following example shows.

Example 5.13. Suppose that the function v is homogeneous of degree $k \geq 1$, is differentiable over Δ_n and satisfies mutual dependency. Let $\mathbf{x}^*, \mathbf{y}^* \in \Delta_n$ be two different global maxima of v in Δ_n. If $\mathbf{x}^* \neq \mathbf{y}^*$, then in particular there exist $i, j \in \Delta_n$ such that $x_i^* > x_j^*$ and $y_i^* < y_j^*$. Under $\mathbf{x}^*$, player j experiences strictly positive regret (he will do much better under y*), and similarly, under y*, player i experiences strictly positive regret.

5.2.1 Non-Differentiable Utility Functions

Theorem 5.12 heavily relies on the fact that v is differentiable; we can, in fact, prove a stronger claim for functions that are not differentiable, but are limits of differentiable functions. We begin by proving a few technical lemmas.

Suppose we are given a sequence of continuous functions $(v_j)_{j=1}^{\infty}$ such that $v_j : \mathbb{R}^n \to \mathbb{R}$ for all j, and v_j uniformly converge to v.

Lemma 5.14. *Suppose that* $\lim_{j \to \infty} \mathbf{x}_j = \mathbf{x}^*$; *then*
$$\lim_{j \to \infty} v_j(\mathbf{x}_j) = v(\mathbf{x}^*).$$

Proof. Given some $\varepsilon > 0$, from uniform convergence we know that there is some $j_0 \in \mathbb{N}$ such that for all $j > j_0$, $|v(\mathbf{x}_j) - v_j(\mathbf{x}_j)| < \frac{\varepsilon}{2}$. Since v_j converge uniformly to v and are continuous, v is continuous as well; therefore, there is some j_1 such that for all $j > j_1$, $|v(\mathbf{x}^*) - v(\mathbf{x}_j)| < \frac{\varepsilon}{2}$. From the triangle inequality we obtain that for all $j > \max\{j_0, j_1\}$, $|v(\mathbf{x}^*) - v_j(\mathbf{x}_j)| < \varepsilon$. □

Lemma 5.15. *Suppose that $\mathbf{x}_j^*$ is a global maximum of v_j over a compact set D for all $j \in \mathbb{N}$, and let $\mathbf{x}^*$ be a unique global maximum of v over D, then* $\lim_{j \to \infty} v_j(\mathbf{x}_j^*) = v(\mathbf{x}^*)$.

Proof. Since D is compact, there is some convergent subsequence $\mathbf{x}_{j_k}^*$ such that $v_j(\mathbf{x}_{j_k})$ converges to a point $v(\mathbf{y}^*)$. Let $\mathbf{x}^*$ be the global maximum of v over D. By uniform continuity we know that
$$\lim_{j \to \infty} v_j(\mathbf{x}^*) = v(\mathbf{x}^*).$$

Therefore $\lim_{k \to \infty} v(\mathbf{x}^*) - v(\mathbf{y}^*)$ equals
$$\lim_{k \to \infty} (v(\mathbf{x}^*) - v_{j_k}(\mathbf{x}^*)) + \lim_{k \to \infty} (v_{j_k}(\mathbf{x}^*) - v(\mathbf{y}^*)),$$
which is at most
$$\lim_{k \to \infty} (v(\mathbf{x}^*) - v_{j_k}(\mathbf{x}^*)) + \lim_{k \to \infty} (v_{j_k}(\mathbf{x}_{j_k}^*) - v(\mathbf{y}^*)).$$

The first limit equals 0 by uniform convergence of v_j, and the second limit equals 0 by Lemma 5.14. This means that $v(\mathbf{x}^*) = v(\mathbf{y}^*)$, hence $\mathbf{y}^*$ is a global maximum of v; since $\mathbf{x}^*$ is a unique global maximum of v over D, we conclude that $\mathbf{y}^* = \mathbf{x}^*$. □

Note that an immediate consequence of Lemma 5.15 is that if v has a unique global maximum $\mathbf{x}^*$ over D then any sequence of maxima of v_j, $\mathbf{x}_j^*$ converges to $\mathbf{x}^*$.

Lemma 5.16. *If $(v_j)_{j=1}^{\infty}$ are homogeneous of degree k and converge uniformly to v, then letting $v_{j,t}(\mathbf{x})$ be the value of v_j at time t under the stationary contract $\mathbf{x}$, we have that $v_{j,t}$ converges uniformly to v_t, and that $u_{i,t,j}(\mathbf{x})$ converges uniformly to $u_{i,t}(\mathbf{x})$, where $u_{i,t,j}$ is the utility of player i under $v_{j,t}$.*

Proof. First, we note that v is k homogeneous:

$$v(\lambda \mathbf{x}) = \lim_{j \to \infty} v_j(\lambda \mathbf{x}) = \lambda^k \lim_{j \to \infty} v_j(\mathbf{x}) = \lambda^k v(\mathbf{x}).$$

Employing Lemma 5.5, we have

$$|v_{j,t}(\mathbf{x}) - v_t(\mathbf{x})| = |v_j(\mathbf{w}(0))^{k^{t-1}} v_j(\mathbf{x})^{H_t} - v(\mathbf{w}(0))^{k^{t-1}} v(\mathbf{x})^{H_t}|$$

which is at most

$$v_j(\mathbf{x})^{H_t} |v_j(\mathbf{w}(0))^{k^{t-1}} - v(\mathbf{w}(0))^{k^{t-1}}|$$
$$+ v(\mathbf{w}(0))^{k^{t-1}} |v_j(\mathbf{x})^{H_t} - v(\mathbf{x})^{H_t}|.$$

Now, taking any $\varepsilon > 0$, there is some j_0 such that $v_j(\mathbf{x}) < 2v(\mathbf{x})$, and both $|v_j(\mathbf{w}(0))^{k^{t-1}} - v(\mathbf{w}(0))^{k^{t-1}}|$, and $|v_j(\mathbf{x})^{H_t} - v(\mathbf{x})^{H_t}|$ are strictly smaller than

$$\frac{\varepsilon}{4 \max\{\max_{\mathbf{x} \in \Delta_n} v(\mathbf{x})^{H_t}, v(\mathbf{w}(0))^{k^{t-1}}\}}.$$

For a *fixed* t, this upper bound does not depend on the value of $\mathbf{x}$. For any $j > j_0$, we have that $|v_{j,t}(\mathbf{x}) - v_t(\mathbf{x})| < \varepsilon$, which concludes the proof. We can repeat the proof for $u_{i,t,j}$, noting that $|u_{i,t,j}(\mathbf{x}) - u_{i,t}(\mathbf{x})| = x_i |v_{t,j}(\mathbf{x}) - v_t(\mathbf{x})|$ which is at most $|v_{t,j}(\mathbf{x}) - v_t(\mathbf{x})|$. □

We are now ready to prove Theorem 5.17.

Theorem 5.17. *Suppose that $(v_j)_{j=1}^{\infty}$ is a sequence of functions that are differentiable, have a unique maximum over Δ_n, and are homogeneous of degree $k \geq 1$; such that v_j converge uniformly to v. Moreover, suppose that v has a unique global maximum over Δ_n. Given a universally optimal stationary contract $\mathbf{x}^* \in \Delta_n$ for v, $\lim_{t \to \infty} p\text{-}reg_i(\mathbf{x}^*, t) = 0$.*

Proof. Take $\mathbf{x}^* \in \arg\max_{\mathbf{x} \in \Delta_n} v(\mathbf{x})$, and let $\mathbf{x}_t^*$ be a global maximum of $u_{i,t}(\mathbf{x})$ over Δ_n. We need to show that for any $\varepsilon > 0$ there is some t_0 such that for all $t > t_0$ we have

$$|u_{i,t}(\mathbf{x}^*) - u_{i,t}(\mathbf{x}_t^*)| < \varepsilon.$$

Let $\mathbf{x}_{t,j}^*$ be a contract in $\arg\max_{\mathbf{x} \in \Delta_n} u_{i,t,j}(\mathbf{x})$ for all t and for all j; let $\mathbf{x}_j^*$ be a contract that is a maximum of v_j for all j. The following inequality holds for all $j \in \mathbb{N}$.

$$\begin{aligned}|u_{i,t}(\mathbf{x}^*) - u_{i,t}(\mathbf{x}_t^*)| &\leq |u_{i,t}(\mathbf{x}^*) - u_{i,t,j}(\mathbf{x}_j^*)| + \\ &\quad |u_{i,t,j}(\mathbf{x}_j^*) - u_{i,t,j}(\mathbf{x}_{t,j}^*)| + \\ &\quad |u_{i,t,j}(\mathbf{x}_{t,j}^*) - u_{i,t}(\mathbf{x}_t^*)|\end{aligned}$$

Since all v_j satisfy Theorem 5.12, we know that there is some t_0 such that for all $t > t_0$, $|u_{i,t,j}(\mathbf{x}_j^*) - u_{i,t,j}(\mathbf{x}_{t,j}^*)| < \frac{\varepsilon}{3}$. Now, according to Lemma 5.15, since $\mathbf{x}_{t,j}^*$ are global maxima of $u_{i,t,j}(\mathbf{x})$ for all j, and $u_{i,t,j}$ converge uniformly to $u_{i,t}$ (by Lemma 5.16), then there is some j_0 such that for all $j > j_0$ we have $|u_{i,t,j}(\mathbf{x}_{t,j}^*) - u_{i,t}(\mathbf{x}_t^*)| < \frac{\varepsilon}{3}$. Finally, since $\mathbf{x}^*$ is a unique global maximum of v, we know from the consequence of Lemma 5.15 that $\lim_{j \to \infty} \mathbf{x}_j^* = \mathbf{x}^*$; therefore, there is some j_0 such that for all $j > j_0$ we have that $|u_{i,t}(\mathbf{x}^*) - u_{i,t,j}(\mathbf{x}_j^*)| < \frac{\varepsilon}{3}$. We conclude that $\lim_{t \to \infty} u_{i,t}(\mathbf{x}_t^*) = u_{i,t}(\mathbf{x}^*)$. □

Theorem 5.12 states that when v is differentiable, strictly concave and homogeneous of degree $k \geq 1$, individual incentives coincide with social welfare in the limit. Theorem 5.17 allows us to drop the differentiability requirement, if we know that v is the uniform limit of a sequence of differentiable functions that satisfy Theorem 5.12.

Contrasting Theorems 5.12 and 5.17 with Lemma 5.3, we obtain a clearer picture of the relation between individual interests and profit maximization in our setting. We stress that even if v satisfies all of the conditions stated in Theorem 5.12, the individually optimal contract for i at time t, $\mathbf{x}_t^*$, must be different from $\mathbf{x}^*$; it must be that the i-th coordinate of $\mathbf{x}_t^*$ is strictly greater than the i-th coordinate of $\mathbf{x}^*$, but the difference between them decreases as t approaches infinity. This can be summarized in the following manner: Lemma 5.3 states that for any t, there is some $\varepsilon_t > 0$ such that $p\text{-}reg_i(\mathbf{x}^*, t) > \varepsilon_t$; Theorems 5.12 and 5.17 show when for any $\varepsilon > 0$ there is some t_ε such that for all $t > t_\varepsilon$, $p\text{-}reg_i(\mathbf{x}^*, t) < \varepsilon$.

5.2.2 Truthful Contracts

Let us observe the following extension to our model. Each player $i \in N$ has some private information m_i, that he reports to the center. Our definition of information is quite general: in the case of data routing, it could represent the player's routing capacity; in the case of company departments, this data can represent internal group assessments or employee evaluations. We denote the set of all possible reports by player i as $\mathcal{M}_i$, and $\mathcal{M} = \mathcal{M}_1 \times \cdots \times \mathcal{M}_n$. Now, the production function v is a function from $\mathcal{M} \times \mathbb{R}_+^n$ to $\mathbb{R}_+$, i.e. the value of v depends both on player resources, and on the private information of the players. A budget allocation mechanism is simply a function from $\mathcal{M}$ to Δ_n, i.e., given some report $\mathbf{m}' \in \mathcal{M}$, the mechanism chooses a point $\mathbf{x}(\mathbf{m}') \in \Delta_n$. If players' true information is given by $\mathbf{m}$, Player i's utility at time t is then $u_{i,t}(\mathbf{m}, \mathbf{x}(\mathbf{m}')) = x_i(\mathbf{m}')v_t(\mathbf{m}, \mathbf{x}(\mathbf{m}'))$; i.e. player reports affect the budget allocation chosen by the mechanism, but the actual values of $\mathbf{m}$ are used. We say that player i is ε-*truthful* at time t if

$$u_{i,t}(\mathbf{x}(\mathbf{m})) \geq u_{i,t}(\mathbf{x}(\mathbf{m}_{-i}, m_i')) - \varepsilon,$$

for all $m_i' \in \mathcal{M}_i$. A mechanism is ε-truthful at time t if for all $\mathbf{m} \in \mathcal{M}$ and all $i \in N$, i is ε-truthful at time t.

Lemma 5.18. *If for all $\mathbf{m} \in \mathcal{M}$ the mechanism chooses $\mathbf{x}(\mathbf{m})$ such that*

$$\lim_{t \to \infty} p\text{-}reg_i(t, \mathbf{m}, \mathbf{x}(\mathbf{m})) = 0,$$

then for all $\varepsilon > 0$, there is some t_0 such that for all $t > t_0$, player i is ε-truthful at time t.

Proof. By reporting m'_i, player i can secure at time t at most $\max_{\mathbf{y} \in \Delta_n} u_{i,t}(\mathbf{m}, \mathbf{y})$, which, for a large enough t, is at most $u_{i,t}(\mathbf{m}, \mathbf{x}(\mathbf{m})) + \varepsilon$, since

$$\lim_{t \to \infty} p\text{-}reg_i(t, \mathbf{m}, \mathbf{x}(\mathbf{m})) = 0.$$

□

Combining Lemma 5.18 with Theorem 5.12 we obtain the following theorem.

Theorem 5.19. *If v is concave, differentiable and homogeneous of degree $k \geq 1$, then for any $\varepsilon > 0$, there is some t_0 such that for all $t > t_0$ choosing the socially optimal contract given $\mathbf{m}$ is an ε-truthful mechanism; if v is homogeneous of degree $k < 1$ then no ε-truthful mechanism exists.*

In order for Theorem 5.19 to hold, it is necessary that the homogeneity and concavity of v do not depend on the parameters that are reported. For example, if v is a CES production function with $r < 1$ (see Section 5.4.1) and players need to report their resource coefficients, then the concavity of v does not depend on the coefficients that the players report. Thus, if all players believe that social welfare is adequately represented by such a function and want to maximize their long-term revenue, they will truthfully report their parameters. However, in the case of Cobb-Douglas production functions (see Section 5.4.2), a player is only incentivized to truthfully report his coefficient if the sum of the coefficients is greater or equal to 1.

5.3 Discounted Returns

It is often the case that future rewards are given with a *discount factor*. Generally speaking, given that the reward at time t is r_t, and a value $0 < \gamma \leq 1$, we define the γ-*discounted reward* at time t as $r_{\gamma,t} = \gamma^{t-1} \cdot r_t$. The parameter γ captures some notion of patience: the higher the value of γ, the more far-sighted the player. We can easily add discount factors to our setting; given a function $v : \mathbb{R}^n \to \mathbb{R}$, an initial resource vector $\mathbf{w}(0)$, and a contract $\mathbf{x} \in \Delta_n$, the γ-*discounted total social welfare* at time t is

$$\text{sw}_{\gamma,t}(\mathbf{x}) = \sum_{h=1}^{t} \gamma^h v_h(\mathbf{x}).$$

In this section we reexamine the results of Section 5.2 when discounts are in place. Assuming again that v is k-homogeneous, the following three lemmas summarize the relationship between the degree of k and the effect on long-term social welfare.

Lemma 5.20. *If v is homogeneous of degree $k < 1$ and $\gamma < 1$ then $\lim_{t \to \infty} \text{sw}_{\gamma,t}(\mathbf{x}) = c$ for some $c \in \mathbb{R}_+$.*

Proof. According to Lemma 5.5, $v_t(\mathbf{x}) = V_1^{k^{t-1}} \cdot v(\mathbf{x})^{H_{t-1}}$, and when $k < 1$, $\lim_{t \to \infty} V_1^{k^{t-1}} \cdot v(\mathbf{x})^{H_{t-1}} = v(\mathbf{x})^{\frac{1}{1-k}}$, which is a constant independent of t. This implies that $\lim_{t \to \infty} \mathrm{sw}_{\gamma,t}(\mathbf{x})$ is indeed a constant if $\gamma < 1$. □

Lemma 5.21. *Suppose v is 1-homogeneous; if $\gamma \cdot v(\mathbf{x}) \geq 1$ then $\lim_{t \to \infty} \mathrm{sw}_{\gamma,t}(\mathbf{x}) = \infty$, and otherwise $\lim_{t \to \infty} \mathrm{sw}_{\gamma,t}(\mathbf{x}) = c$.*

Proof. If v is 1-homogeneous, then $v_t(\mathbf{x}) = V_1 \cdot v(\mathbf{x})^{t-1}$, which means that

$$\mathrm{sw}_{\gamma,t}(\mathbf{x}) = \sum_{h=1}^{t} \gamma^h \cdot V_1 \cdot v(\mathbf{x})^{h-1} = \frac{V_1}{v(\mathbf{x})} \sum_{h=1}^{t} (\gamma \cdot v(\mathbf{x}))^h.$$

The sum $\sum_{h=1}^{t} (\gamma \cdot v(\mathbf{x}))^h$ is divergent if and only if $\gamma \cdot v(\mathbf{x}) \geq 1$, which concludes the proof. □

Lemma 5.22. *Suppose v is homogeneous of degree $k > 1$; if $V_1 \cdot v(\mathbf{x})^{\frac{1}{k-1}} > 1$ then $\lim_{t \to \infty} \mathrm{sw}_{\gamma,t}(\mathbf{x}) = \infty$, otherwise $\lim_{t \to \infty} \mathrm{sw}_{\gamma,t}(\mathbf{x}) = c$.*

Proof. If v is homogeneous of degree $k > 1$, then $v_t(\mathbf{x}) = V_1^{k^{t-1}} \cdot v(\mathbf{x})^{\frac{k^{t-1}-1}{k-1}} = \left(V_1 \cdot v(\mathbf{x})^{\frac{1}{k-1}}\right)^{k^{t-1}} \cdot \frac{1}{v(\mathbf{x})^{\frac{1}{k-1}}}$; thus

$$\mathrm{sw}_{\gamma,t}(\mathbf{x}) = \frac{1}{v(\mathbf{x})^{\frac{1}{k-1}}} \cdot \sum_{h=1}^{t} \gamma^{h-1} \cdot \left(V_1 \cdot v(\mathbf{x})^{\frac{1}{k-1}}\right)^{k^{h-1}},$$

which diverges if and only if $V_1 \cdot v(\mathbf{x}) > 1$. □

The behavior of discounted total welfare at time t strongly depends on the value of k. We stress that the results for $k \neq 1$ hold for any value of $\gamma < 1$. When $k < 1$, $v_t(\mathbf{x})$ converges to a constant, and thus will eventually decay when $\gamma < 1$; this provides even stronger incentives for players not to consider their long-term profits and focus on maximizing their pointwise utility, as shown in Theorem 5.9. When $k > 1$ then the only significant factor is $V_1 \cdot v(\mathbf{x})^{\frac{1}{k-1}}$. In other words, when startup revenue (V_1) is high enough, the increase in social welfare will make any discount factor insignificant in the long-run, i.e. the increase in revenue is so great, that players will always care about their long-term revenue, which makes the eventual agreement of individual incentives and group incentives, as shown in Section 5.2, even more significant. The case when $k = 1$ is the only one where some dependence on γ is exhibited: players care about long-term profits if and only if $\gamma \geq \frac{1}{v(\mathbf{x})}$. In other words, if $v(\mathbf{x}) > 1$ and players are sufficiently patient, then long-term rewards must be considered. For the case where convergence occurs and $k \geq 1$, we have that the value $\gamma^{t-1} v_t(\mathbf{x})$ decreases somewhat proportionally to the rate in which individually optimal contracts converge to the socially optimal contract (approximately k^t when $k > 1$ and t

when $k=1$), which means that while individuals may still wish to renegotiate the selected contract, their incentive to do so decreases as fast as the revenue does. The following theorem captures these observations formally.

Given some $\gamma \in (0,1]$, and a sequence of individually optimal contracts for i, $(\mathbf{x}_t^*)_{t=1}^\infty$, we say that t_γ is an ε-*indifference point*, if it is the first time-step such that for all $t > t_\gamma$, $\gamma^{t-1} u_{i,t}(\mathbf{x}_t^*) < \varepsilon$. In other words, even though player i has a sequence of individually optimal contracts, he would not get more than an ε of discounted revenue any time $t > t_\gamma$. We write $t_\gamma = \infty$ if this never occurs. The following theorem states that if γ is sufficiently large and group revenue is sufficiently high, then player i would care about his profits for a sufficiently long time so that he would eventually agree to the globally optimal contract.

Theorem 5.23. *Given a player $i \in N$, let $(\mathbf{x}_t^*)_{t=1}^\infty$ be a sequence of individually pointwise optimal contracts for i, that converges to some universally pointwise optimal contract $\mathbf{x}^*$. For any $\varepsilon_0 > 0$, let t_0 be the first time so that for all $t \geq t_0$ $|v_t(\mathbf{x}_t^*) - v_t(\mathbf{x}^*)| < \varepsilon_0$. Suppose that for all t we have $v_t(\mathbf{x}^*) \geq c$ for some strictly positive constant c, and let t_γ be the ε_1-indifference point of player i, where $\varepsilon_1 < x_i^* \cdot (c - \varepsilon_0)$; then there is some γ_0 such that for all $\gamma \geq \gamma_0$ we have that $t_\gamma > t_0$.*

Proof. Let us write $\gamma_0 = \left(\frac{\varepsilon_1}{x_i^* \cdot (c-\varepsilon_0)}\right)^{\frac{1}{t_0-1}}$. First, observe that for all $\gamma \geq \gamma_0$, $\gamma^{t_0-1} \geq \frac{\varepsilon_1}{x_i^* \cdot (c-\varepsilon_0)}$, which, by our choice of ε_1, is at most 1. Now, at time t_0, $u_{i,t_0}(\mathbf{x}_{t_0}^*) = x_{i,t_0}^* v_{t_0}(\mathbf{x}_{t_0}^*) \geq x_i^* \cdot (c-\varepsilon_0)$. This immediately implies that at time t_0, $\gamma^{t_0-1} u_{i,t_0}(\mathbf{x}_{t_0}^*) \geq \varepsilon_1$, and in particular, t_γ is strictly greater than t_0. □

Theorem 5.23 states that if players will eventually agree to an optimal contract, and that their revenue does not dwindle down to zero as time goes by, then sufficiently patient players will see the merit of sticking to the optimal contract, which, as noted in the above lemmas, depends on initial allocations and the production function.

To conclude, we observe that in Section 5.2, all convergence results did not depend on the initial resource vector $\mathbf{w}(0)$. When we incorporate discounts, then the interest of players in long-term profits strongly depends on starting conditions.

5.4 Applications

In this section we analyze certain classes of functions to which our results apply.

5.4.1 CES Production Functions

We first explore a family of production functions, known in the literature as *CES production functions* (Constant Elasticity of Substitution). Let $v_r(\mathbf{x}) = c \cdot \left(\frac{1}{n}\sum_{i=1}^n \left(\frac{x_i}{a_i}\right)^r\right)^{\frac{1}{r}}$, where c and $a_1, \ldots, a_n$ are positive constants, and $r \neq 0$. We note that v_r is homogeneous of degree 1 and differentiable for all r. When $r > 1$

v_r is convex, when $r = 1$ it is linear, and when $r < 1$ it is concave. We can in fact show that the optimal contract (for any $r < 1$) is $x_i^* = \frac{\beta_{i,r}}{\sum_{j=1}^n \beta_{j,r}}$, where $\beta_{i,r} = a_i^{\frac{r}{r-1}}$; this is a unique critical point of v_r in the interior of Δ_n, but when $r > 1$, it is a minimum of v_r. Applying our results, we obtain that if $r < 1$, then any sequence of individually optimal contracts necessarily converges to the universally optimal contract described above. In other words, for all strictly concave CES production functions, individual incentives eventually align with social welfare.

It is also known that $\lim_{r \to 0} v_r(\mathbf{x})$ is the Cobb-Douglas production function (Section 5.4.2), and $\lim_{r \to -\infty} v_r(\mathbf{x})$ is the Leontief production function (Section 5.4.3).

5.4.2 Cobb-Douglas Production Functions

A *Cobb-Douglas production function* is a function of the form $V_c(\mathbf{x}) = c \prod_{i=1}^n x_i^{a_i}$. We write $a = \sum_{i=1}^n a_i$. Note that the Cobb-Douglas production function is a-homogeneous: $V_c(\lambda \mathbf{x}) = c \prod_{i=1}^n (\lambda x_i)^{a_i} = \lambda^a V_c(\mathbf{x})$. It is well-known in the economic literature that the maximum of V_c over Δ_n is unique, and equals $(\frac{a_1}{a}, \ldots, \frac{a_n}{a})$; for the sake of completeness, we provide a short proof of this fact.

We wish to solve the equation

$$\max \quad V_c(\mathbf{x}) \qquad (5.1)$$
$$\text{s.t.} \quad \sum_{i=1}^n x_i = 1$$

Note that $\mathbf{x}^*$ is a maximum of $V_c(\mathbf{x})$ in Δ_n if and only if it is a maximum of $\log(V_c(\mathbf{x}))$ over Δ_n (log is the natural logarithm): on one hand, if $\mathbf{x}^*$ is a maximum of $V_c(\mathbf{x})$ then for all $\mathbf{y} \in \Delta_n$ $V_c(\mathbf{y}) < V_c(\mathbf{x}^*) \Rightarrow \log(V_c(\mathbf{y})) < \log(V_c(\mathbf{x}^*))$, since $\log(x)$ is strictly monotone increasing over $\mathbb{R}_+$; on the other hand, if $\log(V_c(\mathbf{x}^*)) > \log(V_c(\mathbf{y}))$, then $V_c(\mathbf{x}^*) > V_c(\mathbf{y})$. Now, $\log(V_c(\mathbf{x})) = \sum_{i=1}^n a_i \log(x_i)$. Thus, a solution to the following equation will also be a solution to Equation (5.1).

$$\max \quad \sum_{i=1}^n a_i \log(x_i) \qquad (5.2)$$
$$\text{s.t.} \quad \sum_{i=1}^n x_i = 1$$

We write $L(\mathbf{x}, \lambda) = \sum_{i=1}^n a_i \log(x_i) + \lambda(1 - \sum_{i=1}^n x_i)$, then $\frac{\partial L}{\partial x_i}(\mathbf{x}) = \frac{a_i}{x_i} - \lambda$ for all $i \in N$, and $\frac{\partial L}{\partial \lambda}(\mathbf{x}) = 1 - \sum_{i=1}^n x_i$. Setting $\frac{\partial L}{\partial x_i} = 0$, we get that $\frac{a_i}{x_i} = \frac{x_j}{a_j}$ for all $i, j \in N$, thus $x_j = \frac{a_j x_i}{a_i}$ for all $i, j \in N$. Since $\sum_{j=1}^n x_j = 1$, we get that for a fixed $i \in N$, $x_i + \sum_{j \neq i} x_j = x_i + \sum_{j \neq i} \frac{a_j x_i}{a_i} = x_i \cdot \frac{a}{a_i} = 1$. To conclude, $x_i = \frac{a_i}{a}$ for all $i \in N$. Since $V_c(\mathbf{x})$ is strictly positive over the interior Δ_n, we conclude that the point we have found must be a global maximum of V_c over Δ_n (there are no other critical points in the interior of Δ_n).

We now turn to measuring pointwise regret in the Cobb-Douglas setting. Since $\mathbf{x}^* = (\frac{a_1}{a}, \ldots, \frac{a_n}{a})$ is a universally optimal stationary contract, and Cobb-Douglas functions are differentiable and homogeneous of degree a, according to Theorem 5.12, we know that for $a \geq 1$, $\lim_{t \to \infty} p\text{-}reg_i(\mathbf{x}^*, t)$ is 0, and when $a < 1$ players exhibit non-vanishing regret from $\mathbf{x}^*$. In the following theorem we provide an explicit formulation of the individually optimal stationary contract for each player when $a < 1$.

Theorem 5.24. *For all $i \in N$, if $a < 1$ then the stationary regret-minimizing contract for player i is $x_i = 1 - \sum_{j \neq i} a_j$ and $x_j = a_j$ for all $j \neq i$.*

Proof. While this claim can be proven using elementary calculus, we present here an alternative proof, which makes use of Theorem 5.9. According to Theorem 5.9, if $(\mathbf{x}_t^*)_{t=1}^\infty$ is a sequence of individually optimal contracts for player i, and $\lim_{t \to \infty} \mathbf{x}_t^* = \mathbf{x}^*$, we have $\frac{\partial V_c}{\partial x_j}(\mathbf{x}^*) = V_c(\mathbf{x}^*)$ for all $j \neq i$, and $\frac{\partial V_c}{\partial x_i}(\mathbf{x}^*) = V_c(\mathbf{x}^*)(1 - \frac{1-a}{x_i^*})$. We note that $\frac{\partial V_c}{\partial x_j}(\mathbf{x}) = \frac{a_j}{x_j} V_c(\mathbf{x})$, so it must be that for all $j \neq i$, $\frac{a_j}{x_j^*} V_c(\mathbf{x}^*) = V_c(\mathbf{x}^*)$, hence $x_j^* = a_j$. For i, we have $\frac{a_i}{x_i^*} V_c(\mathbf{x}^*) = V_c(\mathbf{x}^*)(1 - \frac{1-a}{x_i^*})$, so $x_i^* = 1 - \sum_{j \neq i} a_i$. It can be further shown via an argument similar to that made in Theorem 5.9, tailored to V_c, that indeed there is a unique individually optimal contract for i at any point t, so this sequence converges to this contract. □

5.4.3 Leontief Functions

Consider the function $V_\ell(\mathbf{x}) = c \min_{i \in N} \{\frac{x_i}{a_i}\}$, known as a *Leontief* production function, where c and $(a_i)_{i \in N}$ are all strictly positive constants. We again write $a = \sum_{i=1}^n a_i$. Note that V_ℓ is 1-homogeneous, which implies, by Lemma 5.5 and Corollary 5.6 that the global maximum of V_ℓ is a universally optimal contract. Since V_ℓ is not differentiable, Theorem 5.9 does not apply here. However, we observe that $\lim_{r \to -\infty} v_r(\mathbf{x}) = V_\ell(\mathbf{x})$. If we show that V_ℓ has a unique maximum over Δ_n, then we can apply Theorem 5.17 and show that the unique optimal contract has no regret for any player in the limit. Let $\mathbf{x}^*$ be in $\arg\max V_\ell$; it is clear that $\frac{x_i^*}{a_i} = \frac{x_j^*}{a_j}$ for all $i, j \in N$; combining this fact with $\sum_{i=1}^n x_i^* = 1$, we obtain

$$\sum_{j=1}^n x_j^* \;=\; x_i^* + \frac{x_i^*}{a_i} \sum_{j \neq i} a_j = x_i^* \frac{a}{a_i}$$

which implies that $x_i^* = \frac{a_i}{a}$. Since this optimum is unique, we know that the regret of all players goes to 0 as t grows, according to Theorem 5.17. The Leontief production function exhibits a rather interesting property due to its non-differentiability. Not only is $(\frac{a_1}{a}, \ldots, \frac{a_n}{a})$ universally optimal, it is also an individually pointwise optimal contract for all players for sufficiently large t. Since V_ℓ is not differentiable precisely at the point $(\frac{a_1}{a}, \ldots, \frac{a_n}{a})$, this does not contradict Lemma 5.3, which assumes that v is differentiable.

Theorem 5.25. *The optimal contract is individually pointwise optimal for player i at time t for all $t > \frac{a}{a_i} - 1$. In particular, no player experiences regret at time t for all $t > \max_{i \in N}\{\frac{a}{a_i}\}$.*

Proof. We wish to find the maximum of $u_{i,t}(\mathbf{x}) = x_i v_t(\mathbf{x})$ over Δ_n. We can write $v_t(\mathbf{x}) = c \min\{\frac{x_i}{a_i}, \min_{j \neq i}(\frac{x_j}{a_j})\}$; now, fixing the amount that is allocated to player i to be x, the rest of the players receive $1 - x$, which needs to be distributed in an optimal manner among them. As with finding an optimal contract for v, it must be the case that when player i takes x, an optimal contract for the rest is such that $\frac{x_j}{a_j} = \frac{x_k}{a_k}$ for all $j, k \neq i$. Combining this with the fact that $\sum_{j \neq i} x_j = 1 - x$, we get that

$$\sum_{k \neq i} x_k = x_j + \sum_{k \neq i,j} x_k = x_j + \frac{x_j}{a_j} \sum_{k \neq i,j} a_k = x_j \frac{a - a_i}{a_j};$$

thus, if x is given to i, then the maximal profit is derived when $x_j = (1-x)\frac{a_j}{a-a_i}$. To conclude, the maximal profit to player i at time t is given by finding the maximum of the function $u_{i,t}(x) = xV_1c^{t-1}\min\{\frac{x}{a_i}, \frac{1-x}{a-a_i}\}^{t-1}$. We observe that $\frac{x}{a_i} \leq \frac{1-x}{a-a_i}$ if and only if $x \leq \frac{a_i}{a}$, thus

$$u_t(x) = \begin{cases} V_1 c^{t-1} x \cdot \left(\frac{x}{a_i}\right)^{t-1} & \text{if } x \leq \frac{a_i}{a} \\ V_1 c^{t-1} x \cdot \left(\frac{1-x}{a-a_i}\right)^{t-1} & \text{otherwise.} \end{cases}$$

The function $V_1 c^{t-1} x \cdot \left(\frac{x}{a_i}\right)^{t-1}$ is strictly monotone increasing in $[0, \frac{a_i}{a}]$, so it achieves its maximum at $x^* = \frac{a_i}{a}$. Taking the derivative of $f(x) = V_1 \left(\frac{c}{a-a_i}\right)^{t-1} x(1-x)^{t-1}$, we get

$$\begin{aligned}
\frac{\partial f}{\partial x}(x) &= V_1 \left(\frac{c}{a-a_i}\right)^{t-1} \cdot (1-x)^{t-1} - \\
&\quad V_1 \left(\frac{c}{a-a_i}\right)^{t-1} (t-1)x(1-x)^{t-2} \\
&= V_1 \left(\frac{c}{a-a_i}\right)^{t-1} \cdot (1-x)^{t-2}(1-x-(t-1)x) \\
&= V_1 \left(\frac{c}{a-a_i}\right)^{t-1} \cdot (1-x)^{t-2}(1-tx).
\end{aligned}$$

Thus, the maximum of $f(x)$ at $[\frac{a_i}{a}, 1]$ is at $\max\{\frac{1}{t}, \frac{a_i}{a}\}$. In other words, if $\frac{1}{t} > \frac{a_i}{a}$ (or equivalently, if $t < \frac{a}{a_i}$), then choosing $\frac{1}{t}$ is optimal for player i; otherwise, it is optimal to choose $\frac{a_i}{a}$. $\square$

5.4.4 Network Flow Games

The data center/web routing company example given in the introduction can be thought of as a network flow problem with incentive-driven edges. Suppose we are given a directed, weighted graph $\Gamma = \langle V, E \rangle$, where the weight of the edge $e \in E$ is a positive integer w_e, as well as a pair of nodes $s, t \in V$. We can define the maximum flow game $v_\Gamma : \mathbb{R}^n \to \mathbb{R}$ to be the maximum flow that can be passed through Γ, times some constant value c. Here we assume that $E = \{e_1, \dots, e_n\}$. The constant c can be thought of as the per unit value of the commodity that is passed through the network; in other words, the edges are players, and $v_\Gamma(\mathbf{x})$ is the maximum flow through Γ, given that e has a capacity of $c_e x_e$. The edge e uses the amount of money it has, x_e, to purchase capacity, which is multiplied by the factor c_e. Given $\mathbf{x} \in \mathbb{R}^n_+$, we write $\Gamma(\mathbf{x})$ to be the graph Γ with capacities $c_e x_e$

instead of c_e; thus $\Gamma(\mathbf{1}^n) = \Gamma$, and we indeed assume that $\mathbf{w}(0) = \mathbf{1}^n$. This means that $v_\Gamma(\mathbf{x})$ equals the maximum flow through $\Gamma(\mathbf{x})$. This is a straightforward generalization of the classic network flow cooperative game [Peleg and Sudhölter, 2007].

The first observation we make is that v_Γ is homogeneous of degree 1: changing all edge capacities by a factor of λ results in a change of λ to the maximum flow as well, hence v_Γ is homogeneous of degree 1. Using this fact, we immediately obtain the following result.

Proposition 5.26. *Given a max-flow game v_Γ, it is possible to find a universally optimal stationary contract in polynomial time.*

Proof. Given a valid flow $f = (f_e)_{e \in E}$ from s to t, let $F_{\text{in}}(f, v) = \sum_{e=(u,v)} f_e$ be the total flow into the vertex $v \neq s, t$, and $F_{\text{out}}(f, v) = \sum_{e=(v,u)} f_e$ be the total flow out of v. Since v_Γ is homogeneous of degree 1, according to Lemma 5.5, in order to find a universally optimal stationary contract for v, we need to find a point $\mathbf{x}^* \in \Delta_n$ such that $\mathbf{x}^* \in \arg\max_{\mathbf{x} \in \Delta_n} v_\Gamma(\mathbf{x})$. That is, $\mathbf{x}^*$ is an optimal solution to the following LP:

$$
\begin{aligned}
\max \quad & c \sum_{e \in E} f_e & (5.3) \\
\text{s.t.} \quad & f_e \leq c_e x_e \\
& F_{\text{in}}(f, v) = F_{\text{out}}(f, v) \quad \forall v \in V \setminus \{s, t\} \\
& \sum_{e \in E} x_e = 1
\end{aligned}
$$

LP (5.3) has polynomially many constraints and variables, hence it is solvable in polynomial time.

□

Given an optimal solution to LP (5.3), say $(f^*, \mathbf{x}^*)$, we stress that $(f_e^*)_{e \in E}$ is a max flow of $\Gamma(\mathbf{x}^*)$, not Γ. However, we have that $x_e^* \geq \frac{f_e^*}{w_e}$. For example, if the graph Γ is a directed path, v_Γ reduces to a Leontief function, as described in Section 5.4.3, in which case all players on the path receive some payoff inversely proportional to their capacity. We contrast this result with the canonical core-stable payoff division, which pays only edges that are in the minimum cut (for details, see Maschler et al. [2013]). Paying only the edges in the minimum cut in our setting is clearly not optimal: only the edges in the minimum cut survive the first iteration, and the graph cannot pass any further flow (unless Γ is a degenerate graph where there are no paths of length more than 1 from s to t). It now remains to decide whether universally optimal stationary contracts minimize regret for all players. As the following example shows, the answer is, in general, negative.

Example 5.27. Consider the following graph: there are three vertices, $V = \{s, v, t\}$, and three edges $e_1 = e_2 = (s, v)$ and $e_3 = (v, t)$. The capacity of e_1 and e_3 is a, while the capacity of e_2 is b, where $b < a$. The universally optimal contract in this case is $(\frac{1}{2}, 0, \frac{1}{2})$, i.e. pass the flow through e_1 and e_3, and divide payoffs equally between e_1 and e_3. This clearly has high regret for player 2, who may receive some payoff (at the cost of social welfare) by routing flow via e_2 and e_3 and playing $(0, \frac{1}{2}, \frac{1}{2})$.

In Example 5.27, mutual dependency does not hold; however, as the following example shows, mutual dependency does not guarantee no regret in the limit, when there are several possible maxima.

Example 5.28. Consider again a graph with three vertices $V = \{s, v, t\}$ and three edges $e_1 = e_2 = (s, v)$ and $e_3 = (v, t)$. Now let us set the capacity of e_1 and e_2 to a, and the capacity of e_3 to b, such that $a < b < 2a$. The maximum flow in this network is b, and since $a < b$, all maximal flows must employ all edges. However, since $b < 2a$, there exist maximal flows such that the flow through e_1 is not equal to the flow through e_2, which means that there exist universally optimal contracts such that either e_1 or e_2 have positive regret always.

Let $\mathcal{C}(\Gamma)$ be the set of (s,t) cuts of Γ. Given a cut $C \in \mathcal{C}(\Gamma)$, we can write C as a vector $\mathbf{w}_C$ in $\mathbb{R}^n$, with $\mathbf{w}_C(e) = w_e$ for all $e \in C$, and $\mathbf{w}_C(e) = 0$ otherwise. We let M_Γ be an $|\mathcal{C}(\Gamma)| \times n$ matrix whose rows are the vectors $\mathbf{w}_C$. This means that $v_\Gamma(\mathbf{x})$ can be rewritten as $v_\Gamma(\mathbf{x}) = \min_{C \in \mathcal{C}(\Gamma)} \mathbf{w}_C \cdot \mathbf{x}$. Now, given a vector in $\mathbb{R}^{|\mathcal{C}|}$, we let $f_r(\mathbf{x})$ be a CES production function, much like in Section 5.4.3, i.e. $f_r(\mathbf{x}) = c \left(\sum_{j=1}^{|\mathcal{C}|} \frac{1}{|\mathcal{C}|} x_i^r \right)^{\frac{1}{r}}$ (note that here we assume that $a_1, \ldots, a_n = 1$). We write $v_{\Gamma,r} : \mathbb{R}^n \to \mathbb{R}$ to be $v_{\Gamma,r}(\mathbf{x}) = f_r(M_\Gamma \mathbf{x})$, and noting that $\lim_{r \to -\infty} v_{\Gamma,r}(\mathbf{x}) = v_\Gamma(\mathbf{x})$ with uniform convergence. We note that all $v_{\Gamma,r}$ are homogeneous of degree 1, differentiable in the interior of Δ_n, and when $r < 1$, are strictly concave. Therefore, if v_Γ has a unique global maximum that is in the interior of Δ_n, Theorem 5.17 holds for all $v_{\Gamma,r}$, which means that we can apply the results of Section 5.2 to network flow games.

Theorem 5.29. *If LP (5.3) has a unique solution $(f_e^*, x_e^*)_{e \in E}$, then for any edge e in the network flow game defined by Γ, we have that $\lim_{t \to \infty} p\text{-}reg_e(\mathbf{x}^*) = 0$.*

Chapter 6

Conclusions

One major contribution of this thesis is the notion of arbitration functions, and their role in stabilizing cooperative games with overlapping coalitions. We show that in such games, deciding if a deviation is worthwhile depends to a great extent on the way non-deviators will react to a deviation. In fact, knowing the assumptions one is allowed to make about the reaction of non-deviators to deviation is an absolutely necessary part of the analysis of OCF games, which has not been properly addressed until recently. The reaction of non-deviators to deviation can play a decisive role in the analysis of many strategic interactions; for example, Ackerman and Brânzei [2012] use the concept of arbitration function in the analysis of Nash equilibrium and pairwise equilibrium. More recently, Brânzei et al. [2013] study matchings with externalities; their model can be seen as applying the idea of arbitration functions to matching problems.

In fact, any strategic setting where a deviating set must still interact with non-deviators must make some assumption on the way non-deviators behave when a deviation occurs. In most settings, worst-case behavior is assumed; that is, a notion similar to that of the conservative core is used to assess the profitability of deviation. As we show in Section 2.4.1, this assumption makes OCF games essentially equivalent to non-OCF games when analyzing their stability. This line of reasoning would probably hold for other settings as well. However, a rich underlying structure arises when one does not assume the conservative arbitration function; this type of analysis gives rise to intricate interactions between players. A notion similar to arbitration functions can be easily applied beyond the framework of cooperative games.

We see the arbitration function model as a new paradigm in the study of strategic behavior in multiagent systems. Players can reason about the desirability of a given outcome, and change their behavior in order to affect a new state that will benefit them. However, our work demonstrates that when doing so, it is crucial to address the way that other, non-deviating parties, react. We hope to see this type of reasoning applied in other strategic settings.

There are several models that may benefit the application of the arbitration model. First, market settings see various types of interactions among players. Often, contractual settings clearly define player behavior in market settings. These types of contracts can be modeled as arbitration functions, and thus provide a formal framework for studying market interactions under contracts. Complex reactions to deviation may also appear in social network settings. Players may be more agreeable to forming a link with some players as they provide them with higher utility; however, they may also have different attitudes towards each other, which may or may not be linked to the amount of actual utility they derive from interacting. For example, a group of players may be friends, and will not interact with anyone hurting a member of the group; this type of behavior does not have to depend on the utility derived from the group. By analyzing this type of behavior in the context of social network formation, one may arrive at different types of equilibria, that may be more desirable.

The second major conceptual contribution of our work is the description of a simple and natural dynamic of revenue sharing, where today's revenue is tomorrow's resources. We have analyzed various aspects of stationary profit divisions in this setting, i.e., profit sharing contracts where the share that each player receives at every round is a constant. We have shown that under some conditions, individual and social incentives will coincide, in which case several highly desirable properties arise, including truthful reporting by the players in cases where players possess private information. We believe that the model presented in Chapter 5 is appealing for several reasons. First of all, the results in Chapter 5 highlight the effect of non-myopic behavior on incentives. Players who are completely myopic will not care about their profits in future rounds, and will therefore not see the merit of choosing an outcome that allocates any money to others. Moreover, players who are not sufficiently patient may see the benefits of collaborating with others, but will not do so as their short term gains outweigh (discounted) future rewards. Second, our contribution is one of the first to bridge cooperative game theory and mechanism design. There is a natural connection between the two: consider, for example, a network flow game where players need to report their capacities to a center, who then chooses a payoff division. Does there exist an allocation in the core of the game such that no player gains by misreporting? That is, can one choose an outcome that is both stable and incentive-compatible? In general, the answer to this is negative: there exist network flow games where every core outcome is not incentive compatible [1]. While the revenue sharing scheme proposed in Chapter 5 is not always truthful, we are able to show that it is under some conditions. We hope that other works continue exploring this interesting connection.

6.1 Future Work

Our work provides several perspectives on what are desirable payoff divisions in coalitional settings; there are several avenues of research that can be expanded upon from here. In section 2.4, we have characterized the conservative, sensitive and refined cores, and provided some sufficient conditions for core non-emptiness under these arbitration functions. It would be useful to do this for other natural notions of arbitration, such as the sensitive and optimistic cores.

---

[1] The author thanks Nisarg Shah for highlighting this fact.

One could also greatly benefit from pursuing the "inverse" of this problem; given an OCF game, what is are the "least generous" arbitration functions such that the arbitrated core for that game is not empty? More formally, given a game $\mathcal{G}$, an arbitration function $\mathcal{A}_m$ is *maximal-stable* for $\mathcal{G}$ if $Core(\mathcal{G}, \mathcal{A}_m) \neq \emptyset$, and for all $\mathcal{A}'$ such that $\mathcal{A}' \geq \mathcal{A}_m$ and $\mathcal{A}' \leq \mathcal{A}_m$ we have $Core(\mathcal{G}, \mathcal{A}') = \emptyset$. Simply put, the arbitration function $\mathcal{A}_m$ provides the largest amount of payoffs to deviators without destabilizing $\mathcal{G}$. It would be interesting to even prove that some arbitration function is maximal-stable for a class of OCF games, let alone completely characterizing the set of maximal-stable arbitration functions. These types of results would be especially useful, as they may lead to notions similar to the cost of stability [Bachrach et al., 2009], or minimal taxation schemes [Zick et al., 2013].

In this work, our analysis focuses on the core of cooperative games with overlapping coalitions. This is to an extent a reflection of the state of the art in the study of non-OCF cooperative games; the core and its variants (i.e. the ε-core and the least core) have been studied in greater detail than other solution concepts (barring, perhaps, the Shapley value). Further analysis of the various solution concepts, and their geometric relations to one another, would greatly improve our understanding of OCF games.

From the algorithmic perspective, many open problems arise. For example, it is well-established that finding $\mathcal{A}$-stable outcomes is computationally infeasible for many classes of games. It might be possible to find *approximately stable* outcomes much faster; in particular, we can employ a notion similar to the cost of stability in classic cooperative games [Bachrach et al., 2009] and adapt it to OCF games. Briefly, given a classic cooperative game with an empty core, the cost of stability is the extra subsidy required in order to stabilize the game. The lower the subsidy required, the better. In fact, the algorithms presented in Chapter 3 can be slightly modified in order to obtain the cost of stability of an OCF game, for a given arbitration function. Recently, Meir et al. [2013] show that when the Myerson graph of a cooperative game has a treewidth of k, the cost of stability of the game is at most $k + 1$. Moreover, Meir et al. [2013] present some poly-time algorithms for finding $(k + 1)$-stable payoff divisions in polynomial time for monotone, simple games (i.e., one needs to pay at most $(k + 1)$ times more than the value of the optimal coalition structure in order to stabilize the game). While there is no reason to assume that finding approximately stable outcomes for OCF games will be any easier —it is, in fact, likely to be much harder due to the complexities detailed in Chapter 3— bounds on the cost of stability could possibly be shown for more complex arbitration functions. In general, the more lenient the arbitration function, the higher the cost of stability. Here, perhaps, lies a possible way of mitigating not only the theoretical bounds on the subsidy required in order to stabilize a game, but also the computational complexity of finding an approximately stable outcome. By employing a stricter reaction to deviation, one can achieve more stable outcomes, and thus reduce the cost of stability.

Other natural approaches to breaching the computational barrier are finding $\mathcal{A}$-stable outcomes for restricted classes of games. The results we show for LBGs (Section 3.6) are a first step in this direction, and many other natural classes of OCF games can be studied for their computational complexity. Computing outcomes in the arbitrated nucleolus and bargaining set is likely to present a serious computational challenge, as finding nucleolus outcomes is already a complex task, even for rather simple classes of games [Elkind and Pasechnik, 2009, Chen et al., 2012].

Finally, expanding the iterated profit sharing model presented in Chapter 5 is an important step in understanding dynamics of iterated cooperation. In this thesis, we assume that the same contract is used at every time-step. This assumption can be sometimes justified: it is straightforward, easy to implement and understand, and is often used in real-world settings (for example, a long-term budget set by a large corporation dictates revenue shares for several rounds). However, we can easily model dynamic environments, where a new contract is negotiated at every time step. Finding a sequence of optimal contracts is a challenging task; however, preliminary studies show that in some settings, it is possible to find optimal dynamic contracts in polynomial time; moreover, dynamic contracts tend to converge to a stationary contracts eventually. In the dynamic contracts setting, it is not immediately clear how one would judge the desirability of a contract to an individual. Simply extending our notion of regret to the dynamic setting will not work: the best dynamic contract at time t for player i is to have players maximize utility until time $t-1$, and then give all of the revenue to the i at time t; therefore, an alternative notion of regret needs to be utilized.

The structure of incentives in the iterated revenue division model we propose assumes non-myopic players; this leads to an alternative solution concept for the one-shot game: assign payoffs in the one-shot game as if it were an iterated game. The relation between this solution concept and other solution concepts —either those introduced here, such as the refined core, or classic solution concepts such as the Shapley value and the core— would certainly provide us with a better understanding of what constitutes a good revenue division in a collaborative setting.

Publications Related to this Thesis

1. Y. Zick, Y. Bachrach, I. Kash and P. Key. Non-Myopic Negotiators See What's Best. *Proceedings of the 24th international joint conference on artificial intelligence (IJCAI'15)*, 2015.

2. Y. Zick, E. Markakis and E. Elkind. Stability via Convexity and LP-duality in OCF Games. *Proceedings of the 26th AAAI Conference on AI (AAAI-12)*, pages 1506–1502, 2012

3. Y. Zick, G. Chalkiadakis and E. Elkind. Overlapping Coalition Formation Games: Charting the Tractability Frontier. *Proceedings of the 11th international joint conference on Autonomous agents and multiagent systems (AAMAS-12)*, pages 787–794, 2012.

4. Y. Zick and E. Elkind. Arbitrators in Overlapping Coalition Formation Games. *Proceedings of the 10th international joint conference on Autonomous agents and multiagent systems (AAMAS-11)*, pages 55–62, 2011.

Bibliography

M. Ackerman and S. Brânzei. Research quality, fairness, and authorship order. *CoRR*, abs/1208.3391, 2012.

E. Anshelevich and M. Hoefer. Contribution games in social networks. In *Proceedings of the 18th European Symposium on Algorithms (ESA-10)*, pages 158–169, 2010.

T. Arnold and U. Schwalbe. Dynamic coalition formation and the core. *Journal of economic behavior & organization*, 49(3):363–380, 2002.

J.P. Aubin. Cooperative fuzzy games. *Mathematics of Operations Research*, 6(1): 1–13, 1981.

R.J. Aumann and J.H. Drèze. Cooperative games with coalition structures. *International Journal of Game Theory*, 3:217–237, 1974.

Y. Bachrach, E. Elkind, R. Meir, D. Pasechnik, M. Zuckerman, J. Rothe, and J. Rosenschein. The cost of stability in coalitional games. In *Proceedings of the 2nd International Symposium on Algorithmic Game Theory (SAGT-09)*, pages 122–134, 2009.

O.N. Bondareva. Some applications of linear programming methods to the theory of cooperative games. *Problemy kibernetiki*, 10:119–139, 1963.

R. Brafman, C. Domshlak, Y. Engel, and M. Tennenholtz. Transferable utility planning games. In *Proceedings of the 24th AAAI Conference on AI (AAAI-10)*, pages 709–714, 2010.

R. Brânzei, D. Dimitrov, and S. Tijs. *Models in cooperative game theory*. Springer, 2005.

S. Brânzei, T. Michalak, T. Rahwan, K. Larson, and N. R. Jennings. Matchings with externalities and attitudes. In *Proceedings of the 12th International Conference on Autonomous Agents and Multi-Agent Systems (AAMAS-13)*, pages 295–302, 2013.

G. Chalkiadakis, E. Elkind, E. Markakis, M. Polukarov, and N.R. Jennings. Cooperative games with overlapping coalitions. *Journal of Artificial Intelligence Research*, 39:179–216, 2010.

G. Chalkiadakis, E. Elkind, and M. Wooldridge. *Computational Aspects of Cooperative Game Theory.* Morgan and Claypool, 2011.

N. Chen, P. Lu, and H. Zhang. Computing the nucleolus of matching, cover and clique games. In *Proceedings of the 26th AAAI Conference on AI (AAAI-12)*, 2012.

T.H. Cormen, C.E. Leiserson, R.L. Rivest, and C. Stein. *Introduction To Algorithms.* MIT Press, 2001.

B. Courcelle. The monadic second-order logic of graphs. i. recognizable sets of finite graphs. *Information and Computation*, 85:12–75, 1990.

T.M. Cover. Universal portfolios. *Mathematical Finance*, 1(1):1–29, 1991.

V. D. Dang, R. K. Dash, A. Rogers, and N. R. Jennings. Overlapping coalition formation for efficient data fusion in multi-sensor networks. In *Proceedings of the 21st AAAI Conference on AI (AAAI-06)*, pages 635–640, 2006.

M. Davis and M. Maschler. Existence of stable payoff configurations for cooperative games. *Bulletin of the American Mathematical Society*, 69:106–108, 1963.

G. Demange. On group stability in hierarchies and networks. *Journal of Political Economy*, 112(4):754–778, 2004.

X. Deng, T. Ibaraki, and H. Nagamochi. Algorithmic aspects of the core of combinatorial optimization games. *Mathematics of Operations Research*, 24(3):751–766, 1999.

E. Elkind and D. Pasechnik. Computing the nucleolus of weighted voting games. In *Proceedings of the twentieth Annual ACM-SIAM Symposium on Discrete Algorithms (SODA-09)*, pages 327–335, 2009.

E. Elkind, D. Pasechnick, and Y. Zick. Dynamic weighted voting games. In *Proceedings of the 12th International Conference on Autonomous Agents and Multiagent Systems (AAMAS-13)*, pages 515–522, 2013.

E. Even-dar, Y. Mansour, and U. Nadav. On the convergence of regret minimization dynamics in concave games. In *Proceedings of the 41st annual ACM symposium on Theory of computing (STOC-09)*, pages 523–532, 2009.

M. R. Garey and D. S. Johnson. *Computers and Intractibility.* W. H. Freeman and Company, 1979.

E. Gofer and Y. Mansour. Pricing exotic derivatives using regret minimization. In *Algorithmic Game Theory*, pages 266–277. Springer, 2011.

G. Greco, E. Malizia, L. Palopoli, and F. Scarcello. On the complexity of core, kernel, and bargaining set. *Artificial Intelligence*, 175(12-13):1877–1910, 2011.

M. O. Jackson. A survey of models of network formation—stability and efficiency. In G. Demange and M. Wooders, editors, *Group Formation in Economics: Networks, Clubs and Coalitions*, chapter 1. Cambridge University Press, 2003.

K. Jain and M. Mahdian. Cost sharing. In N. Nisan, T. Roughgarden, E. Tardas, and V.V. Vazirani, editors, *Algorithmic Game Theory*, chapter 15, pages 383–408. Cambridge University Press, 2007.

Ehud Lehrer and Marco Scarsini. On the core of dynamic cooperative games. *Dynamic Games and Applications*, 3(3):359–373, 2013.

C.F. Lin and S.L. Hu. Multi-task overlapping coalition parallel formation algorithm. In *Proceedings of the 6th International Joint Conference of Autonomous Agents and Multiagent Systems (AAMAS-07)*, page 211, 2007.

E. Markakis and A. Saberi. On the core of the multicommodity flow game. *Decision support systems*, 39(1):3–10, 2005.

M. Maschler, B. Peleg, and L. S. Shapley. Geometric properties of the kernel, nucleolus, and related solution concepts. *Mathematics of Operations Research*, 4(4):303–338, 1979.

M. Maschler, E. Solan, and S. Zamir. *Game Theory*. Cambridge University Press, Hebrew University of Jerusalem, Israel, 2013.

R. Meir, Y. Zick, E. Elkind, and J. S. Rosenschein. Bounding the cost of stability in games over interaction networks. In *Proceedings of the 27th AAAI Conference on AI (AAAI-13)*, pages 690–696, 2013.

R.B. Myerson. Graphs and cooperation in games. *Mathematics of Operations Research*, 2(3):225–229, 1977.

N. Nisan, T. Roughgarden, É. Tardos, and V.V. Vazirani, editors. *Algorithmic Game Theory*. Cambridge University Press, 2007.

B. Peleg and P. Sudhölter. *Introduction to the Theory of Cooperative Games*, volume 34 of *Theory and Decision Library. Series C: Game Theory, Mathematical Programming and Operations Research*. Springer, Berlin, second edition, 2007.

N. Robertson and P.D Seymour. Graph minors. III. planar tree-width. *Journal of Combinatorial Theory, Series B*, 36(1):49 – 64, 1984.

Tuomas Sandholm, Kate Larson, Martin Andersson, Onn Shehory, and Fernando Tohmé. Coalition structure generation with worst case guarantees. *Artificial Intelligence*, 111(1):209–238, 1999.

D. Schmeidler. The nucleolus of a characteristic function game. *SIAM Journal on Applied Mathematics*, 17:1163–1170, 1969. ISSN 0036-1399.

L.S. Shapley. A value for n-person games. In *Contributions to the Theory of Games, vol. 2*, Annals of Mathematics Studies, no. 28, pages 307–317. Princeton University Press, 1953.

L.S. Shapley. On balanced sets and cores. *Naval Research Logistics Quarterly*, 14(4):453–460, 1967.

L.S. Shapley. Cores of convex games. *International Journal of Game Theory*, 1: 11–26, 1971.

O. Shehory and S. Kraus. Formation of overlapping coalitions for precedence-ordered task-execution among autonomous agents. In *Proceedings of the 2nd International Conference on Multi-Agent Systems (ICMAS-96)*, pages 330–337, 1996.

V.V. Vazirani. *Approximation algorithms*. Springer Verlag, 2001.

T. Voice, M. Polukarov, and N.R. Jennings. Coalition structure generation over graphs. *Journal of Artificial Intelligence Research*, 45:165–196, 2012.

E. Winter. The shapley value. *Handbook of Game Theory with Economic Applications*, 3:2025–2054, 2002.

H.P. Young. Monotonic solutions of cooperative games. *International Journal of Game Theory*, 14(2):65–72, 1985.

Y. Zick and E. Elkind. Arbitrators in overlapping coalition formation games. In *Proceedings of the 10th international joint conference on Autonomous agents and multiagent systems (AAMAS-11)*, pages 55–62, 2011.

Y. Zick, M. Polukarov, and N. R. Jennings. Taxation and stability in cooperative games. In *Proceedings of the 12th International Conference on Autonomous Agents and Multiagent Systems (AAMAS-13)*, pages 523–530, 2013.

M. Zinkevich. Online convex programming and generalized infinitesimal gradient ascent. In *Proceedings of the 20th International Conference on Machine Learning (ICML-03)*, 2003.

www.ingramcontent.com/pod-product-compliance
Lightning Source LLC
Chambersburg PA
CBHW080921170526
45158CB00008B/2185